Probability, Statistics and Time

Probability, Statistics and Time

A collection of essays

M. S. BARTLETT, F.R.S.

Emeritus Professor of Biomathematics
University of Oxford

LONDON
CHAPMAN AND HALL

A Halsted Press Book
John Wiley & Sons, Inc., New York

First published 1975
by Chapman and Hall Ltd
11 New Fetter Lane, London EC4P 4EE

Typesetting by Santype (Coldtype Division) Ltd,
Salisbury, Wiltshire
Printed in Great Britain by
University Printing House, Cambridge

ISBN 0 412 14150 7

Distributed in the U.S.A. by Halsted Press,
a Division of John Wiley & Sons, Inc., New York

Library of Congress Cataloging in Publication Data

Bartlett, Maurice Stevenson,
Probability, Statistics and Time

(Monographs on Applied Probability and Statistics)
1. Probabilities – Addresses, essays, lectures.
2. Mathematical statistics – Addresses, essays, lectures.
I. Title.
QA273.18.B37 519 75-24171
ISBN 0-470-05466-2

Preface

Some years ago when I assembled a number of general articles and lectures on probability and statistics, their publication (*Essays in Probability and Statistics*, Methuen, London, 1962) received a somewhat better reception than I had been led to expect of such a miscellany. I am consequently tempted to risk publishing this second collection, the title I have given it (taken from the first lecture) seeming to me to indicate a coherence in my articles which my publishers might otherwise be inclined to query.

As in the first collection, the articles are reprinted chronologically, usually without comment. One exception is the third, not previously published and differing from the original spoken version both slightly where indicated in the text and by the addition of an Appendix.

I apologize for the inevitable limitations due to date, and also for any occasional repetition of the discussion (e.g. on Bayesian methods in statistical inference). In particular, readers technically interested in the classification and use of nearest-neighbour models, a topic raised in Appendix II of the fourth article, should also refer to my monograph *The Statistical Analysis of Spatial Pattern* (Chapman and Hall, London, 1976), where a much more up-to-date account of these models will be found, and, incidentally, a further emphasis, if one is needed, of the common statistical theory of physics and biology.

March 1975 M.S.B.

Contents

page

1. **Probability, Statistics and Time** (Inaugural lecture at University College, London, on 15 May, 1961) 1

2. **R. A. Fisher and the last Fifty Years of Statistical Methodology** (The first R. A. Fisher Memorial Lecture to be given in the United States, on 29 December, 1964 at Chicago; reprinted from *J. Amer. Statist. Ass.*, **60** (1965), 395–409.) 21

3. **The Paradox of Probability in Physics** (Based on a talk with this title given to a Philosophy of Science Group at University College, London, on 22 May 1967) 36

4. **Inference and Stochastic Processes** (Presidential Address to the Royal Statistical Society on 21 June, 1967; reprinted from the Journal of the Society, **A130**, 457–77) 51

5. **Biomathematics** (Inaugural lecture in the University of Oxford, on 28 May 1968) 72

6. **When is Inference *Statistical* inference?** (Invited paper given at a Symposium on the Foundations of Statistical Inference at Waterloo, Canada, April, 1970; reprinted from the Proceedings of the Symposium, 20–31) 98

7. **Epidemics** (Invited article, first published in *Statistics: A Guide to the Unknown.* Tanur, Judith M., Mosteller, F. *et al.*, (Ed). San Francisco: Holden-Day) 111

8. **Equations and Models of Population Change** (Invited paper given at a Conference on the Mathematical Theory of the Dynamics of Biological Populations, Oxford, September, 1972, subsequently edited by R. W. Hiorns and myself, and published by the Academic Press, 1973) 124

9. **Some historical remarks and recollections on multivariate analysis** (Invited paper given at a Conference on Multivariate Analysis and its Applications at Hull, April, 1973; reprinted from *Sankhya,* **36B** (1974), 107–114.) 141

Acknowledgments

Acknowledgments are gratefully made for permission to reprint the articles noted, as follows:

1. University College, London, and the publishers, H. K. Lewis & Co. Ltd. London.

2. The American Statistical Association.

4. The Royal Statistical Society, London.

5. The Clarendon Press, Oxford.

6. From *Foundations of Statistical Inference* A Symposium edited by V. P. Godambe and D. A. Sprott. Copyright © 1971 by Holt, Rinehart and Winston of Canada, Limited. Reprinted by permission of Holt, Rinehart and Winston of Canada, Limited.

7. From *Statistics: A Guide to the Unknown* edited by Judith M. Tanur, Fred Mosteller *et al.* Reprinted by permission of Holden-Day, Inc., Publishers, San Francisco.

8. The Institute of Mathematics and its Applications, and the Academic Press publishers of *The Mathematical Theory of the Dynamics of Biological Populations.*

9. The Editorial Committee of *Sankhya.*

Probability, statistics and time

Reprinted by permission of
H. K. Lewis & Co Ltd., London

I THINK I AM RIGHT IN CLAIMING TO BE GIVING THE first Inaugural Lecture in Statistics at this College. At least, while I feel very honoured at following what I might term the Pearsonian dynasty (father and son), I will remind you that, after the retirement of Karl Pearson from the Galton Chair in 1933, Egon Pearson became Head of a new Department of Statistics, separate from the Department of Eugenics, but the Chair of Statistics was not established till later; and apparently the first Statistics Professor escaped the obligation which the custom of Inaugurals imposes on the majority of us new Professors. My responsibility must have begun to weigh heavily with me, for I have found myself somewhat preoccupied over the content and title of my lecture. My first title 'Statistics as an academic subject' I discarded as too dull, although it did imply one quandary which confronts university statisticians—that statistics as a subject is, like many other live subjects, very difficult to fit into the Procrustean bed of a departmental syllabus. My second discarded title 'Is statistics respectable?' I rejected as not respectable enough for an inaugural lecture, though some of you who heard Professor Cramér's lectures at the end of last term will, like me, have been intrigued at his own reference to the lack of respectability which the subject of probability, so much the theoretical basis of statistics, possessed only a generation ago. The third of my discarded titles, 'The status of statistics', was a compromise that I might have adopted, for I felt strongly about the contrast, between the vital rôle which statistics occupies both in day-to-day affairs and in all aspects of modern science, and the neglect with which the subject, in spite of its growing recognition, has still been treated in some quarters. There are of course special features about statistics as a subject that create difficulties—for example, what is its relation with mathematics; how among the vast range of applications does one find its central discipline; will its very breadth of applications make it become too 'pure', like geometry; what about the interminable controversies over statistical induction? However, I was sceptical of the value of talk which might merely smack of propaganda, and felt that a survey of some of the developments in recent years should make a better case for statistics than any more direct plea.

Naturally enough, I shall refer to developments that I have myself been interested in; and this will provide one explanation of the word 'time' in my final title, for it is statistical or stochastic processes in time that have especially interested me for some years, as indeed they have interested many others. Later on in my lecture, as my specific examples may seem rather ordinary, I shall risk the luxury of some comments on rather more general scientific issues.

Some of you may notice the similarity between my title and that of the collection of lectures by the late Richard von Mises, namely (in the English translation) *Probability, Statistics and Truth*. The difference is of course deliberate; I cannot say, though I hope to have time to discuss, what is time, let alone 'what is truth'. Nevertheless, on referring to Mises, provided one looks beyond the cumbersome and now obsolete mathematical apparatus that he attempted to make a basis for his frequency theory of probability, one is struck by the modern relevance of much of his discussion. I should feel well satisfied with this lecture if I could say when I end, like Mises,

> I think I have succeeded in demonstrating the correctness of the thesis indicated in the title and in the introduction. . . . This was that if we started with a logically clear concept of probability, based on experience, and using arguments which are usually called statistical, we can discover the truth in many fields of human interest.[1]

II MODERN THEORETICAL TRENDS

(1) *Historical and present*

If I want to indicate the trends in statistical thought over the last twenty-five years or so, it is rather natural for me to think back to

[1] loc. cit., pp. 306-7.

my own days as a University student at Cambridge, attending the first course on mathematical statistics at that University given in 1931-32 by the late Dr J. Wishart. Wishart had the double qualifications of having studied both under Karl Pearson at this College and under R. A. Fisher, then at Rothamsted; and his lectures indicated the influence of these two great statisticians by containing, on the one hand, a sound study of theoretical distributions and their numerical comparison with data and, on the other, an introduction to the design and analysis of agricultural or other statistical experiments.

My introduction to probability theory was more garbled, for English statisticians had always treated abstract Continental mathematics with some indifference; and, to complicate matters more, at Cambridge a school of subjectivists, including J. M. Keynes and F. P. Ramsey, and following with Harold Jeffreys, were right on our doorstep.

Lastly, I had the good fortune to meet and hear Udny Yule, although he had officially retired; from him I first met the problem of making sense of correlations obtained from time-series, a problem which did not fit in with any of the statistics I had been taught.

To round off this educational reminiscence, I spent a year here as assistant lecturer under Egon Pearson; at the same time I encountered statistical problems in genetics from R. A. Fisher and J. B. S. Haldane, both recently appointed Professors in the College.

I hope I am not labouring unduly over this brief autobiography, but one's study and research in a subject is very dependent on the leads given by a few individuals, and particular aspects can stagnate for a generation, or over a whole continent, if the time and place are not ready. In a few years the exact theory of statistical inference, developed, especially on the estimation side, by Fisher, was to be supplemented on the side of testing hypotheses (however much Fisher himself queried the importance of their contribution) by J. Neyman and E. S. Pearson, working together in those days at University College.

Jeffreys continued to press the claims of inverse probability; but my own suspicion of theories that seemed over-introspective made me feel, rather vaguely, that the incontrovertible contributions to statistics were going to be in the realm of new methods and techniques and, in particular, in breaking away from the rather static outlook that persisted at that time. Looking back, any contributions of my own before the war on this theme were extremely peripheral, but as a partial excuse I can only cite the situation that prevailed. No comprehensive theory of time-series capable of coping with Yule's problems appeared to exist; evolutionary probability in genetics was making progress, but not theoretically in any very systematic manner; sequential sampling was still to be invented. Even physics, which might have been expected to provide some useful hints, was comparatively stagnant in this respect. Thermodynamics rested on a purely equilibrium theory of statistical mechanics; quantum theory, in spite of its great developments, rested on a probability foundation which appeared to some of us paradoxical (it still does); the statistical theory of turbulence was still in its infancy. In the analysis of time-series and in genetics, one slight obstacle may have been the concentration on discrete-time systems, which often tended to generate rather complicated algebra. Two more serious reasons for the slow progress, however, were, firstly, that in our own country two epoch-making papers on random processes in continuous time, by A. G. McKendrick, one in 1914 entitled 'Studies on the theory of continuous probabilities with special reference to its bearing on natural phenomena of a progressive nature', and one in 1926 on 'Applications of mathematics to medical problems', had been quite overlooked; secondly, that the Continental work on the theory of stochastic processes, especially the Russian work by A. N. Kolmogorov on the general theory and by A. Khintchine and E. Slutsky on the theory of time-series, had still to permeate through to this country. A beginning with a much-needed pooling of knowledge occurred just before the war with a contribution by H. Wold in which he brought together the work of the Russian and British schools on time-series; and, returning once more to my personal education, I learnt something more of the Continental work from J. E. Moyal, who had during the war migrated to this country from Paris.

After the war progress on the theory and applications of stochastic processes was rapid; in this country, for example, the 1949 Symposium on Stochastic Processes arranged by the Royal Statistical Society, with D. G. Kendall, J. E. Moyal, and myself as the three speakers, aroused considerable interest. The developments both here and abroad have grown so much that it is difficult now to keep abreast of recent contributions: as one illustration of this the International Statistical Institute has appointed an international team of experts, coordinated by Wold, to compile a bibliography on the theory of time-series, and associated developments in the theory of stochastic processes. Apart from the specific subject of the statistical analysis of time-series, much of this progress represented either pure theory or the construction of stochastic models as an aid to interpreting random phenomena, and one of my own general aims was to keep all this development properly integrated with the more classical methods of statistical analysis. I shall presently refer to one or two simple examples of stochastic processes which may help to indicate their connexion with problems of inference.

It would be wrong of me, while concentrating on this vast development, which I think is exciting and a great challenge to young statisticians, who can begin to advance from it rather than to it, to ignore general progress in statistical thought, and in particular developments in the fundamental principles of statistical inference. However, at the risk of being biased, I shall complete my remarks on stochastic processes first, because I think there is there less room for argument, and, because of this and of the direct aim of much of the theory in describing real phenomena, more hope of permanence.

(2) *Examples of random or stochastic processes*

There is hardly any need for me here to expound generally on the subject of stochastic processes, especially as many of the more technically minded among you would have been present at Professor Cramér's lectures last term. In considering particular fields, I might remind you of M. G. Kendall's references in his Presidential address to the Royal Statistical Society to stochastic models and processes in the social sciences. I might call attention also to a recent address by Neyman (1960) to the American Statistical Association, in which he noted the importance of stochastic processes for modern statisticians, and discussed applications both in medicine and in astronomy. I, myself, with others, have in the last few years been struggling with applications in ecology and epidemiology; however, as I have summarized some of this work recently in a small monograph, I do not propose to burden you with it again now.

Another recent development worth mentioning is represented by the study of the flow of road traffic as a stochastic process, a development to which Alan Miller has made a valuable contribution. Dr Miller's impending move from the Statistics Department at University College to the Department of Highway and Traffic Engineering at Birmingham compels me to combine regrets at his departure with good wishes to him in his new post, a post which will encourage him to continue with his researches. The broad picture of traffic flow that emerges is that of a completely random process under low traffic density gradually changing to a queueing process as the density increases, with the unfortunately familiar end-point of complete congestion after a certain stage. I have purposely used the words *queueing* and *congestion* to emphasize the connexion with an important general topic in the theory of stochastic processes studied lately, so-called *congestion* and *queueing* theory.

My first example will be a much more specific problem that arose in some correspondence from Bolivia with Dr M. H. Nederlof, to whom I am in consequence indebted. Ten years ago I put forward an extension[1] of the well-known chi-square test—originally developed by Karl Pearson to test the goodness of fit of any hypothetical probabilistic model to observed numerical frequencies obtained from a sample of independent observations—to testing the goodness of fit of Markov chains, that is, sequences of categories where the probability of any one occurring depended on the category that had just occurred. Dr Nederlof pointed out to me a useful application of my procedure for testing the tendency for some sedimentary sequences investigated by geologists to give cyclic patterns of succession. He cited a sequence quoted in the second edition of the book *Sedimentary Rocks* by F. J. Pettijohn; this sequence is summarized in Table I in terms of five main

[1] This is described in my book *An Introduction to Stochastic Processes* (1955).

lithologies: sandstone (S), clay (K), coal or carbonaceous shale (C), limestone (L), and marine-type shale (M).

TABLE I

	S	K	C	L	M	Total
S	0	1	1	2	6	10
K	8	0	2	3	1	14
C	0	10	0	1	0	11
L	0	3	4	0	7	14
M	1	0	4	9	0	14
Total	9	14	11	15	14	63

Table I is a sort of contingency table, but each entry represents the frequency with which the particular lithology at the top of each column is followed by the particular lithology given for each row (for example, C follows K ten times). It will be noticed that every diagonal entry is zero, and this corresponds to the absence of two consecutive lithologies being identical (for they would then be regarded as one). The particular Markov chain under test is the comparatively simple one of a completely random sequence apart from the restriction just mentioned, so that the transition probabilities appropriate to every column are fairly obvious functions of the marginal probabilities, which can be estimated from the data. I won't weary you with the details of the calculation, but we find a chi-square of 49·81 with 11 degrees of freedom (for s categories, the degrees of freedom are $s^2 - 3s + 1$), a value which is very highly significant, and indicative of the need for an alternative hypothesis such as the cyclic tendency (which is here S K C L M). For technical completeness I should note that the rather small numbers would render the exact significance level a little 'woolly' but it may be shown that pooling of categories to increase the frequencies per cell is permissible if desired in this example (in general this is not true for Markov chains).

As another example let me refer to the work of an amateur speleologist, Dr R. L. Curl, who wondered whether any reasonable statistical theory could account for the observed distributional features of caves in regard to their lengths, numbers of entrances, and so on. One of the most well-known stochastic processes or models nowadays, though it is as well to remember that its theory is less than twenty-five years old, is the so-called simple birth-and-death process. In this model the chance of each individual giving rise to a new individual is assumed to be λdt in the infinitesimal time-interval dt, and the chance of dying μdt. Another model, which I have called the emigration-immigration process (and applies, incidentally, to the number of occupied lines in a telephone exchange), identifies dying with emigration, but in place of births allows immigrations from outside, with a chance νdt of a new immigrant individual.

To apply this latter model to caves, Dr Curl assumed that the chance of an entrance to a cave disappearing was proportional to the number of entrances, and the chance of an entrance arising was proportional to the length of the cave. Thus if we consider entrances per cave, we have ν replaced by $\nu' s$, where s is the cave length. The equilibrium distribution of number of entrances is then known to be a Poisson distribution with mean ν/μ or, in the present application, $\nu' s/\mu$. We thus see that on this model not only should the distribution of number of cave entrances be Poisson for given s, but its mean should be linearly related to s. This is the unadjusted mean, and in the analysis one must modify the mean to allow for the failure to observe caves with no entrances at all! In practice it is necessary to group over s to some extent, but I think there is an advantage in keeping different groups non-overlapping, instead of integrating, as Dr Curl did, over cumulative aggregates of caves longer than particular s such as 50, 100, 500, and 1,000 feet. However, a rough check along these lines from the recorded data confirmed Dr Curl's estimate of 0·0003 per foot for the mean $\nu' s/\mu$, an estimate which he had found reasonably consistent not only over different cave aggregates within West Virginia and Pennsylvania, but also from one of these States to the other.

In a later paper Dr Curl proceeded with some success to discuss a stochastic model for cave length. While he does not claim any final validity for his models without further investigation, he states the case for them adequately enough by remarking: 'The alternative to a stochastic model is to maintain that every cave is unique and that no processes may be identified as acting in common upon all caves.'

My acquaintance with this example dates from meeting Dr Curl at Berkeley last summer, and I do hope Dr Curl is able, as he plans, to be with us at University College from next October for a year and to study these and other examples of stochastic processes further.

III MORE FUNDAMENTAL QUESTIONS

(1) *Probability and inference*

Even although in these examples we have seen a reasonable blend of theory and observation, of, if you like, probability and statistics, the rôle of the theory has been, while non-classical in its basis of random processes and stochastic models evolving in time, fairly direct in conception. There are deeper and often more controversial issues associated with the use and interpretation of probability in science—some of these affect the statistician more directly than others, but because of his own continual use of probability theory as a tool and technique he will in any case hardly remain indifferent to its use in general.

The first issue is how subjective or objective should probability be. In statistics there has been some revival of the subjective approach, which I mentioned earlier was not unfamiliar to English statisticians, but which received some further impetus from the development in the United States of what is called 'decision theory' and in particular from an American book by L. J. Savage on the *Foundations of Statistics*. Such contributions, which have caused statisticians to re-examine the extent to which their current methods of analysis are logical and optimal, have been extremely valuable; sometimes, however, as I have emphasized on other occasions, there has been a neglect of the distinction between the statistical conception of some phenomena and the scientific study of all phenomena. This has been coupled with an over-simplification of the problem of induction in terms of degrees of belief, or possibly of 'utilities', which many of us regard as dangerous if used too glibly.

The argument is really part of an even more general issue, which is unlikely to be resolved for many years yet; this is the rôle of the observer in science, the question of what he can observe and what he must invent. In Karl Pearson's day the essential function of the scientist, both as observer and as theorist, was less well understood, so that his remarks must have sounded even more provocative than they do now when he wrote:

> Law in the scientific sense is . . . essentially a product of the human mind and has no meaning apart from man. It owes its existence to the creative power of his intellect. There is more meaning in the statement that man gives laws to Nature than in its converse that Nature gives laws to man.[1]

Nevertheless, while we are fully conscious that every scientific concept is hypothetical and man-made (and this of course applies in particular to statistical concepts), this does not lessen the value of these concepts nor make us relinquish the idea that they represent in some sense the world around us. Schrödinger (1958) expressed it in his own vivid style when, speaking of this problem in connexion with quantum theory, he said:

> Yet it ought to be possible, so we believe, to form in our mind of the physical object an idea . . . that contains in some way everything that *could be* observed in some way or other by any observer, and not only the record of what *has been* observed simultaneously in a particular case. I mean precisely what someone (was it not Ernst Mach?) has called the completion of facts in thought. . . . We prefer to grasp the shape of a solid by visualizing it in three dimensional space instead of by a set of perspective drawings, even though the eye can at any moment only perceive one perspective view. Everyday life is based every minute on 'completion in thought' since we rely on the continued existence of objects while they are not observed by anyone; e.g. we surmise the nocturnal preservation of our portfolio and its contents, locked up in a drawer at night and taken out in the morning.

Schrödinger summed up this thesis by saying:

> The wide-spread attitude that the claim for an objective description of physical reality must be given up, is rejected on the ground that the so-called external world is built up exclusively of elements of the single minds, and is characterized as what is common to all, recognized by every healthy and sane person. Hence the demand for a non-subjective description is inevitable, of course, without prejudice whether it be deterministic or otherwise.

[1] *The Grammar of Science* (1937), p. 77.

(2) *Probability and quantum mechanics*

The last phrase in Schrödinger's remarks: 'without prejudice whether it be deterministic or otherwise' refers of course to the controversy among physicists on the ultimate goal for physical laws, deterministic or statistical. The statistician feels a sneaking satisfaction that he is not after all the only one who indulges in professional controversy, though perhaps a little intrigued that the controversy seems to involve his particular kind of concepts. In the end, after trying to understand what the fuss is about, and finding that Einstein, de Broglie, and Bohm appeared to have argued in favour of determinism, in opposition to the orthodox non-deterministic and statistical view represented by the so-called Copenhagen school, he is a little bewildered that something essentially a matter of conception could be any more than a subject for discussion on, not what is the true, but what is the most fruitful, approach.

I ought to explain here that statisticians are accustomed to postulating randomness or chance as a basic conception against which purpose and pattern can be detected; it is, however, fruitless to discuss whether such a basic *really* exists. For this reason I cannot agree with Landé in his recent book on quantum physics when (arguing against determinism) he says (page 13):

> 'Statistical co-operation' of individual events in a statistical ensemble, whether ordinary or quantal, does not admit of a deterministic explanation.

We do not, of course, require it to, but it might; the use of pseudo-random numbers by electronic computers is a well-known illustration of this. What I (in common with many others—including not only Landé, but some of its founders like de Broglie, Einstein, and Schrödinger) have always felt to be unsatisfactory about quantum mechanics is the *ad hoc* introduction of probability in the theoretical formulation; it is this, in effect, that there has been such a lot of argument about. I will remind you that one of the basic concepts in quantum mechanics is Schrödinger's wave-function ψ, which is not only complex-valued, but which is related to a probability via its squared modulus. The properties of ψ automatically cope with the well-known Uncertainty Principle between position and momentum, though only in a relativistic formulation does the corresponding uncertainty between energy and time appear symmetrically. Statisticians and electrical engineers are familiar with an analogous uncertainty between time and frequency[1] in the analysis of time-series, and this obviously suggests the query: can a frequency ν be associated with an energy E? Physicists appeal to the relation $E=h\nu$, where h is Planck's constant, but quite apart from the qualms expressed by Schrödinger (1958) about this relation, it is at least arguable that the frequency ν is as fundamental in it as the energy E. I can therefore sympathize with (though I am sceptical of) the proposals by Bohm and de Broglie for a return to the interpretation of ψ in terms of real (deterministic) waves; I do not think these proposals will be rebutted until the statistical approach has been put on a more rational basis. Interesting attempts have been made by various writers, but none of these attempts so far has, to my knowledge, been wholly successful or very useful technically.

For example, Landé keeps to a particle formulation, whereas it is the particle, and its associated energy E, which seem to be becoming the nebulous concepts. Let me refer again to time-series theory, which tells us that the quantization of a frequency ν arises automatically for circularly-defined series—for, if you will allow me to call it this, periodic 'time' (more precisely in a physical context, for the angle variables which appear in the dynamics of bound systems). A probabilistic approach via *random fields* thus has the more promising start of including naturally two of the features of quantum phenomena which were once regarded as most paradoxical and empirical—the Uncertainty Principle and quantization. This switch to fields is of course not new; the real professionals in this subject have been immersed in fields for quite a while.[2] However, I am not sure that what probabilists and what physicists mean here by *fields* are quite synonymous, and in any case it is the old probabilistic interpretation in terms of particles that we lay public still get fobbed off with. It would seem to me useful at this stage to make quite clear to us where, if anywhere, the particle

[1] Frequency in the sense of the reciprocal of a wave period.

[2] Cf. Professor Bohm's remarks in *Causality and Chance in Modern Physics*, p. 119.

aspect is unequivocal—certainly discreteness and discontinuity are not very relevant.

Here I must leave this fascinating problem of probability in quantum mechanics, as I would like to turn to its function in the theory of information.

(3) *The concept of information*

Information theory as technically defined nowadays refers to a theory first developed in detail in connection with electrical communication theory by C. Shannon and others, but recognized from the beginning as having wider implications as a conceptual tool. From its origin it was probably most familiar at first to electrical engineers, but its more general and its essentially statistical content made it a natural adjunct to the parts of probability theory hitherto studied by the statistician. This is recognized, for example, in an advertisement for a mathematical statistician from which I quote:

> Applicants should possess a degree in statistics or mathematics, and should if possible be able to show evidence of an interest in some specialized aspect of the subject such as, for example, decision theory, information theory or stochastic processes.

It has not, I think, been recognized sufficiently in some of the recent conferences on information theory, to which mathematical statisticians *per se* have not always been invited.

The close connection of the information concept with probability is emphasized by its technical definition in relation to an *ensemble* or population, and indeed, it may usefully be defined (cf. Good (1950), Barnard (1951)) as $-\log p$ (a simple and direct measure of uncertainty which is reduced when the event with probability p has occurred), although the more orthodox definition is the 'average information' $-\Sigma p \log p$, averaged over the various possibilities or states that may occur. It is also possible to extend this definition to partial or relative information, in relation to a change of *ensembles* or distributions from one to another. With this extended definition of $-\log p/p'$, where p' relates to the new ensemble, the information can be positive or negative, and as the logarithm of a probability ratio will look familiar to statisticians, although it should be stressed that the probabilities refer to fully specified distributions, and the likelihood ratio of the statistician (made use of so extensively by Neyman and E. S. Pearson) only enters if the probabilities p and p' are interpreted as dependent on different hypotheses H and H'. For example, if p' is near p, differing only in regard to a single unknown parameter θ, then

$$E_H\{-\log p/p'\} \sim -\tfrac{1}{2}E_H\left\{-\frac{\partial^2 \log p}{\partial\theta^2}\right\}(\Delta\theta)^2 = -\tfrac{1}{2}I(\theta)(\Delta\theta)^2,$$

$$E_{H'}\{-\log p/p'\} \sim \tfrac{1}{2}I(\theta)(\Delta\theta)^2,$$

where $I(\theta)$ is R. A. Fisher's information function, under conditions for which this function exists.

Formally, the concept of information in Shannon's sense can be employed more directly for inferring the value of θ. To take the simplest case shorn of inessentials, if we make use of Bayes's theorem to infer the value of a parameter θ_r which can take one of only k discrete values, then our prior probability distribution about θ_r will be modified by our data to a posterior probability distribution. If we measure the uncertainty in each such distribution by $-\Sigma p \log p$, we could in general expect the uncertainty to be reduced, but we can easily think of an example where the data would contradict our *a priori* notions and make us less certain than before. This seems to me to stress the subjective or personal element in prior probabilities used in this way, and my own view is that the only way to eliminate this element would be deliberately to employ a *convention* that prior distributions are to be maximized with respect to uncertainty. In the present example this would imply assuming a uniform prior distribution for θ_r, and ensure that information was always gained from a sample of data; it is somewhat reminiscent of arguments used by Jeffreys in recent years for standardizing prior distributions, but I think it important to realize that such conventions weaken any claim that these methods are the only rational ones possible.

Whether or not the information concept in this sense finds any permanent place in statistical inference, there is no doubts of its potential value in two very important scientific fields, biology and physics. This claim in respect to biology is exemplified by the

Symposium on Information Theory in Biology held in Tennessee in 1956; and while we must be careful not to confuse the general function of new concepts in stimulating further research with the particular one of making a particular branch or aspect of a science more precise and unified, the use of the information concept in discussing capacities of nerve fibres transmitting messages to the brain, or coding genetic information for realization in the developed organism, should be sufficient demonstration of its quantitative value. As another illustration of the trend to more explicit and precise uses of the information concept in biology, we may consider the familiar saying that life has evolved to a high degree of organization, that in contrast to the ultimate degradation of dead matter, living organisms function by reducing uncertainty, that the significant feature of their relation with their environment is not their absorption of energy (vital of course as this is), but their absorption of negative entropy. An attempt to measure the rate of accumulation of genetic information in evolution due to natural selection has recently been made by Kimura (1961), who points out that a statement by R. A. Fisher that 'natural selection is a mechanism for generating an exceedingly high degree of improbability' indicates how the increase in genetic information may be quantitatively measured. While his estimate is still to be regarded as provisional in character, it is interesting that Kimura arrives at an amount, accumulated in the last 500 million years up to man, of the order of 10^8 'bits'[1], compared with something of the order of 10^{10} bits estimated as available in the diploid human chromosome set. He suggests that part of the difference, in so far as it is real, should be put down to some redundancy in the genetic coding mechanism.

With regard to physics, I have already mentioned 'negative entropy' as a synonym for information, and this is in fact the link. Again we have the danger of imprecise analysis, and the occurrence of a similar probabilistic formula for information and physical entropy does not by itself justify any identification of these concepts. Nevertheless, physical entropy is a statistical measure of disorganization or uncertainty, and information in this context a

[1] A 'bit' is (apart from sign) a measure of the uncertainty arising from two equally likely alternatives.

reduction of uncertainty, so that the possibility of the link is evident enough. To my mind one of the most convincing demonstrations for the need of this link lies in the resolution of the paradox of Maxwell's demon, who circumvented the Second Law of Thermodynamics and the inevitable increase in entropy by letting only fast molecules move from one gas chamber to another through a trap-door.

It has been pointed out by Rosenfeld (1955) that Clausius in 1879 went some way to explaining the paradox by realizing that the demon was hardly human in being able to discern individual atomic processes, but logically the paradox remains unless we grant that such discernment, while *in principle feasible*, at the same time creates further uncertainty or entropy at least equal (on the average) to the information gained. That this is so emerges from a detailed discussion of the problem by various writers such as Szilard, Gabor, and Brillouin (as described in Brillouin's book).

(4) *The rôle of time*

I might have noted in my remarks on quantum theory that, whether or not time is sometimes cyclic, it appears in that theory in a geometrical rôle, reminiscent of time in special relativity, and not in any way synonymous with our idea of time as implying evolution and irreversible change. It is usually suggested that this latter rôle must be related to the increase of physical entropy, but when we remember that entropy is defined statistically in terms of uncertainty we realize not only that evolutionary time itself then becomes statistical, but that there are a host of further points to be sorted out.

Let me try to list these:

(*a*) In the early days of statistical mechanics, at the end of the last century, Maxwell's paradox was not the only one raised. Two others were Loschmidt's reversibility paradox, in which the reversibility of microscopic processes appeared to contradict the Second Law, and Zermelo's recurrence paradox, in which the cyclical behaviour of finite dynamic systems again contravened the Second Law. It should be emphasized that, while these paradoxes were formulated in terms of deterministic dynamics, they were not immediately dissipated by the advent either of quantum theory or of the idea of statistical processes. For I have just reminded you that time in quantum mechanics is geometrical and reversible; and stationary statistical processes based on microscopic reversible processes are themselves still reversible and recurrent.

The explanations of the paradoxes are based, in the first place, on the difference between absolute and conditional probabilities, and in the second, on the theory of recurrence times. The apparent irreversibility of a system is due to its being started from an initial state a long way removed from the more typical states in equilibrium and the apparent non-recurrence of such a state to the inordinately long recurrence time needed before such a state will return.

(*b*) So far so good—but this conclusion applies to a system of reasonable size. We conclude that microscopic phenomena have no *intrinsic* time-direction, at least if this can only be defined in relation to internal entropy increase (cf. Bartlett, 1956). This is consistent with theoretical formulations in recent years of sub-atomic phenomena involving time-reversals.

(*c*) We have also to notice that while the entropy of our given system will increase with external or given time, this relation is not reciprocal, for, if we first choose our time, a rare state in our stationary process will just as likely be being approached as being departed from. It was argued by Reichenbach that we should consider a whole set of 'branch systems' beginning with low entropy, the time-direction being then defined in terms of increasing entropy for the majority. But this hardly disposes of the problem, for such branch systems can be regarded as part of a larger system which must have been started off in a low entropy configuration.

E. A. Milne avoided this dilemma by treating the direction of time as psychological in relation to which other phenomena are classified. This is in one sense reasonable, but I believe the more fundamental view to be that human beings are subject to physical laws as well as inanimate objects. Our sense of time then follows from a point effectively made by Eddington that the fact that intelligent beings are asking these questions inevitably restricts the starting point of our over-all system to that of low enough entropy. Incidentally, this remark has some bearing on the plausibility of the Cosmological Principle, which asserts that observers anywhere in the universe will be presented with the same over-all (that is, statistical) picture. This is plausible for observers taken at random, but becomes less so if the observers have to be sufficiently highly selected.

(*d*) One final point to remember is that physical entropy cannot be defined for a completely specified isolated system; it is either immersed in a larger system or contains hidden motions which are left unspecified. We cannot therefore define entropy for the entire universe (a point realized by E. A. Milne, though not by Jeans); this excludes from our present discussion the question whether the *entire* universe is stationary or evolutionary, though the question may make sense for the *observable* part.

The summary conclusion, and hardly a new one, is that entropy relations impose both a lower and an upper limit to systems with a time-direction. There is in principle, as distinct from detail, not even perhaps anything very startling in this conclusion, for may I recall Karl Pearson's own remarks on space and time which he summarized.

> Space and time are not realities of the phenomenal world, but the models under which we perceive things apart. They are not infinitely large or infinitely divisible, but are essentially limited by the contents of our perception.[1]

[1] *The Grammar of Science* (1937), p. 163

References

Barnard, G. A. (1951), 'The theory of information', *J. R. statist. Soc.* B, 13, 46.

Bartlett, M. S. (1955), *An Introduction to Stochastic Processes* Cambridge University Press. 'Irreversibility and statistical theory', talk given to a Philosophy of Science Group, Manchester, 1956. In *Essays on Probability and Statistics.* (1961), Methuen, London. (1960), *Stochastic Population Models in Ecology and Epidemiology*, Methuen, London.

Bohm, D. (1957), *Causality and Chance in Modern Physics*, Routledge, London.

Brillouin, L. (1956), *Science and Information Theory,* Academic Press, New York.

Broglie, L. de. (1960), *Non-Linear Wave Mechanics*, Amsterdam.

Curl, R. L. (1958), 'A statistical theory of cave entrance evolution', *Bull. nat. speleol. Soc.*, **20**, 9.

(1960), 'Stochastic models of cavern development', *Ibid.*, **22**, 66.

Fabri, E. (1959), 'Time reversal and complex numbers in quantum theory', *Nuovo Cim.*, (10) **13**, 326.

Good, I. J. (1950), *Probability and the Weighing of Evidence*, MITP, London.

Kendall, M. G. (1961), 'Natural law in the social sciences', *J. R. statist. Soc.* A, **124**, 1.

Kimura, M. (1961), 'Natural selection as the process of accumulating information in adaptive evolution', *Genetical Research,* **2**, 127.

Landé, A. (1960), *From Dualism to Unity in Quantum Physics*, Cambridge University Press.

Miller, A. (1961), A queueing model for road traffic flow', *J. R. statist. Soc.* B, **23**, 64.

Middleton, D. (1960), *Statistical Communication Theory*, New York.

Mises, R. von. (1939), *Probability, Statistics and Truth*, London.

Neyman, J. (1960), 'Indeterminism in science and new demands on statisticians', *J. Amer. statist. Ass.,* **55**, 625.

Pearson, Karl. (1892), *The Grammar of Science*, Everyman ed. (1937).

Pettijohn, F. J. (1957), *Sedimentary Rocks*, 2nd ed. Har-Row, New York.

Reichenbach, H. (1956), *The Direction of Time*, University of California Press, Berkeley.

Rosenfeld, L. (1955), 'On the foundations of statistical thermodynamics', *Acta phys. polon.*, **14**, 3.

(1961), 'Foundations of quantum theory and complementarity', *Nature*, **190**, 384.

Savage, L. J. (1954), *Foundations of Statistics*, Dover, New York.

Schrödinger, E. (1958), 'Might perhaps energy be a merely statistical concept?', *Nuovo Cim.*, **9**, 162.

Wanless, H. R. and Weller, J. M. (1932), 'Correlation and extent of Pennsylvanian cyclothems', *Bull. geol. Soc. Amer.*, **43**, 1003.
Yockey, H. P. (editor), (1958), *Information Theory in Biology*, Pergamon Press, London.

R.A. Fisher and the last fifty years of statistical methodology

Reprinted from the *Journal of the American Statistical Association* (June 1965)
Vol. **60**, pp. 395–409

1. OPENING REMARKS

FIRSTLY may I express my appreciation of this invitation to address this conference. I feel particularly honoured at being the first to give this R. A. Fisher Memorial Lecture, perhaps partly because I have always tried to combine my profound admiration of his scientific achievements with a reluctance to be blinded by their brilliance; and it could be a matter of opinion how far such an attempt at detachment qualified me in particular to discuss his work. I understand that there was no obligation in this lecture to consider Fisher's work, but it seemed rather odd if in this first memorial lecture one did not take the opportunity to do so; and a convenient way of doing this, in view of Fisher's great influence on the development of statistics during the last fifty years, was to try to survey this period, with particular reference to the position Fisher occupied.

Let me make it clear that I am going to concentrate on Fisher's contributions to statistics. It is well-known that his work in genetics was of comparable status; it is largely represented by his book *The Genetical Theory of Natural Selection*, though in his subsequent work his further association with ecological and experimental studies in evolutionary genetics, and his share in the development of studies in the human blood groups, might especially be recalled. Let me also stress that, in the best and fullest sense of the phrase, I am thinking of Fisher as a working statistician. My reference in the title of this lecture to statistical methodology implies a deliberate emphasis on statistical method and technique (as distinct, say, from mathematical probability or the finer details of mathematical rigour), as the field where Fisher was for so long pre-eminent. Of course, as we shall see, we cannot separate statistical methodology from the theory of statistical inference: but it is sometimes advisable, when we find ourselves getting over-excited about the more controversial points of inductive logic, to remember the extent of the common and permanent body of statistical techniques now available to us, techniques which if they are to have proper scientific status should be as far as possible independent of the particular philosophy of the statistician practicing them.

2. FISHER AND STATISTICAL INFERENCE

Fisher's contributions to statistics began with a paper in 1912 advocating the method of maximum likelihood for fitting frequency curves, although the first paper of substance was his 1915 paper in *Biometrika* on the sampling distribution of the correlation coefficient. The stream of statistical papers which followed, especially after he had been appointed as statistician at Rothamsted, can perhaps be divided into three main lines, though of course all interconnected and exemplifying Fisher's general outlook, which I will comment on further in a moment. The first of these three lines consisted of the spate of solutions of exact sampling distribution problems, for which Fisher's geometrical approach proved so powerful. The second was the development of a more general and self-contained set of principles of statistical inference, especially for estimation problems. The third was the emergence of a precise technique of experimental design and analysis.

Now the second line did involve the systematic study of the large-sample or asymptotic properties of maximum likelihood estimates of unknown parameters, a study which is obviously classifiable under asymptotic or large-sample theory. Nevertheless, Fisher was also introducing even here concepts applicable to small samples such as information, likelihood and sufficiency; and by and large, his aim to provide a logically and mathematically precise theory of statistical inference in all its aspects seems fairly clear-cut. His ultimate degree of success I will come to later, but there is no question of the tremendous and immediate impact that so many of his results had because of their practical importance and value in statistical methodology. His consideration of small-sample theory, especially of exact small-sample distributions, was of course not new. The distribution for '*t*' had been correctly conjectured by W. S. Gosset ('Student') in 1908, and the problem of the correlation coefficient distribution was, while previously unsolved, already posed. In fact Fisher at the beginning of his 1915 *Biometrika* paper refers not only to Student's 1908 paper on the '*t*' distribution, but to a subsequent paper of Student's in the same volume on the distribution of the correlation coefficient where the exact result, in the null case, was also correctly conjectured. Nevertheless, it seems fair to say that statistical methods were mostly only available, at least in regard to the assessment of their accuracy, in the large-sample sense; and it will be recalled that the use of Karl Pearson's goodness of fit criterion χ^2 was necessarily rather fuzzy until the degrees of freedom controversy had been resolved by Fisher. This comprehensive tidying-up of exact distributions was, however, only a concomitant of his assault on the principles of statistical inference in general. Here of course he was on much more controversial ground. The theory of probability, on which any statistical principles must hinge, was still considered shaky both logically and mathematically. The mathematical foundations of probability theory were hardly satisfactorily formulated until 1933 by Kolmogorov. When

Fisher wrote his 1925 paper on estimation theory, he was obliged to insert an introductory paragraph on his definition of mathematical probability. This, in terms of infinite hypothetical populations, was a bit crude. Yet it was the minor issue, as no one doubted that the rules of probability could be successfully applied to statistical phenomena. The logical question was much more fundamental. How far could one isolate the inference problems in statistics from the inference problems in science, or indeed in everything?

It is to Fisher's credit that he succeeded in getting some way. He tried to pose problems of analysis as the reduction and simplification of statistical data. He put forward his well-known concept of amount of information* in estimation theory, such that information might be lost, but never gained (only extracted), by analysis. His concept has been of great practical value, especially in large-sample theory. His attempt to extend its relevance to small samples, by considering the case of an accumulation of small samples, was ingenious, but still strictly tied to large samples; it was rather curious that he did not notice the result later discovered by various workers and known as the Cramér-Rao inequality. But this result is still tied to mean square error, an arbitrary criterion in small samples, so that in any case small-sample theory is left more open. On the reduction of data, Fisher's concept of sufficiency was relevant both for estimation and testing purposes; though (i) we do not always have sufficiency (ii) even if we do, we still have to make an inference from our sufficient statistic. While Fisher was strongly critical of inverse probability methods (the so-called Bayesian approach), and rightly emphasized the relevance of the likelihood function as far as the data were concerned, he did not always make it clear exactly how much he was claiming—on the fundamental issue of induction itself I always found his writings extremely cryptic. Moreover if others attempted to expand or develop incomplete parts of the theory of statistical inference, such as Neyman and Pearson with their general theory of testing statistical hypotheses, he was downright rude! Yet the power of a test as introduced by these authors was a valuable tool in studying the statisticsl properties of tests in general; and often a very salutory reminder that a negative result in a significance test from a small sample might not imply as much as experimenters versed in Fisherian methods were sometimes inclined to believe.

It now seems convenient to group tests broadly into two classes; those that are merely 'goodness of fit' tests, involving perhaps many degrees of freedom, and not necessarily very sensitive to particular departures from the null hypothesis i.e. hypothesis under test; and those that refer specifically to one parameter, or at most a few parameters, and often better formulated as problems of confidence intervals (or regions) for the parameter (or parameters) concerned. Such intervals, if efficiently calculated, indicate automatically the accuracy with which a parameter is estimated, and include in effect a test of an entire range of possible values of the parameter. Unfortunately, this theory of confidence intervals, as developed by Neyman, is no longer synonymous with the theory of fiducial intervals, as developed by Fisher. I say 'unfortunately' because I regard this divergence as a regrettable red herring in the more per-

* $I(\theta) = E\{(dL/d\theta)^2\}$, where $L = \log p(S|\theta)$ is the logarithm of the probability of a random sample S when an unknown parameter has true value θ.

manent controversy between Bayesian and non-Bayesian methods. If I linger over it for a while, this is not to imply I want to over-emphasize its significance in the evolution of statistical methodology, but because Fisher, who felt strongly about it, did not argue fairly with his critics on this issue and as one of the founder members of this particular group I consider I have some right to comment. I suggested a moment ago, as indeed did John Tukey at the 1963 I.S.I. meeting at Ottawa,* that these arguments in the higher flights of statistical logic ought not to make much difference to our final conclusions in any particular statistical analysis (assuming of course that the analysis is based on a reasonable amount of statistical data). Why then do we bother with any of these more controversial issues? Are they not practically irrelevant to the development of statistical methodology and technique? Fortunately yes, to a large extent; but in a deeper sense, all of us tend to be affected in our more immediate tasks by our mental attitudes and general philosophies. We cannot therefore expect to divorce statistical methodology entirely from our philosophy of induction; I do hope, however, that we can keep a sense of perspective. I have no expectation of resolving the wider issues by anything I can say in this lecture, and do not propose to do more than remind you of them before I finish. First, however, I shall digress for a moment on the narrower issue of fiducial probability.

3. FIDUCIAL INFERENCE

Let me recall the situation in 1930. Fisher had just published a paper in the Proceedings of the Cambridge Philosophical Society entitled "Inverse probability." It was in this brief note that Fisher defined a 'fiducial interval' for an unknown parameter θ, with a known 'fiducial probability' that the true value of the parameter was covered by the interval. Discussing the case of an unknown correlation coefficient ρ, he said ' . . . if we take a number of samples . . . from the same or different populations and for each calculate the fiducial 5 per cent value for ρ, then in 5 per cent of cases the true value of ρ will be less than the value we have found' (loc. cit., p. 535). Fisher's wording clearly implied, and it was the implication accepted at the time, that the fiducial interval calculated in terms of the sample statistic r was a random interval, and fiducial probability a statistical probability with the usual frequency interpretation. Such an interpretation was even current at the time outside professional statistical circles, for compare the remarks made by Eddington (who would have been in touch with Fisher's work) in his *New Pathways in Science* (1935, p. 126): 'We can never be sure of particular inferences; therefore we should aim at a system of inference that will give conclusions of which in the long run not more than a stated proportion, say $1/q$, will be wrong.'

At this date fiducial intervals were a particular class of 'confidence intervals,' particular because Fisher restricted his theory to exact intervals calculated from density functions, and to sufficient statistics. From the standpoint of inductive logic it must be stressed that the solutions can only be formulated in terms of a *hypothetical* statistical framework. Nevertheless, as a formal alternative to any Bayes solution it was a precise, ingenious and useful statistical technique.

* See p. 941–4, Linnik *et al.* (1964).

It might be noticed that Fisher's original example was not too happy, as nuisance parameters were strictly involved, and Fisher never himself examined in detail the problem of sufficiency in the case of more than one parameter. In the notorious extension to the difference in means of two samples from normal populations with different variances, I personally believe that Fisher made a straightforward mistake due to thinking of s^2 as a sufficient statistic for σ^2, and hence* s_1^2/s_2^2 as a sufficient statistic for σ_1^2/σ_2^2 (if this had been true, the Behrens-Fisher solution of the problem would have been acceptable on the orthodox frequency interpretation, and no controversy would have arisen). As circumstantial evidence I might point to Fisher's expression† for the information on a mean from a normal sample of size N with unknown variance viz. $N^2/[(N+2)s^2]$, an expression which has an infinite average in repeated samples from the same population for $N \leq 3$ and which, as an information formula, I regard as meaningless. In my original critical paper (1936a) on the Behrens-Fisher problem, I tentatively proposed the information formula

$$\frac{(N-2)^2}{(N-1)s^2 + N(\bar{x} - m)^2},$$

with an average in repeated sampling from the same population of $(N-2)/\sigma^2$. The fireworks that followed the publication of my paper unfortunately annihilated, at least for some time, any further discussion of this separate problem. This to my mind is a pity as after nearly 30 years this problem is still to my knowledge not entirely resolved.

The trouble arises through the factorization of the probability function into two exact factors which nevertheless involve the unknown mean m. It follows that the usual dropping or cancellation of the differential element in the likelihood function or ratio is not immediate, and this creates some ambiguity in the proper definition of the appropriate likelihood function. The definition I adopted led to the splitting of the log likelihood equation for several samples $i = 1 \cdots n$ from populations with common mean m but differing variances σ_i^2 into two portions as follows:

$$\frac{dL}{dm} = \frac{dL'}{dm} + \frac{dL''}{dm} = 0$$

where

$$\left\{\begin{aligned} \frac{dL}{dm} &= \sum_{i=1}^{n} \frac{N_i(\bar{x}_i - m)}{\sigma_i^2}, \\ \frac{dL'}{dm} &= \sum_{i=1}^{n} \frac{(N_i - 2)N_i(\bar{x}_i - m)}{(N_i - 1)s_i^2 + N_i(\bar{x}_i - m)^2}. \end{aligned}\right.$$

These equations are consistent with the theory of information, but led to the paradox that samples of 2 or less contained no usable information. I do not think this is necessarily absurd when we remember that a very technical defi-

* This claim, which I consider an error, is repeated by Yates (1964, p. 347).
† See Fisher (1935b). §74.

nition of information is being used. The mean $\bar{x}_i$ still contains the amount of information N_i/σ_i^2, whether we know σ_i^2 or not; the question here is what fraction of this can be combined with that from other samples in an equation providing an estimate with precise and optimum accuracy. (For further discussion relating to this problem, see Neyman & Scott (1948) and James (1959).)

Coming next to the difference in means problem itself, I pointed out at the time that with two observations $x_{i(1)}$, $x_{i(2)}$ for each of two samples ($i=1, 2$) either of the statistics

$$t' = \frac{x_{1(1)} + x_{1(2)} - x_{2(1)} - x_{2(2)}}{|x_{1(1)} - x_{1(2)} + x_{2(1)} - x_{2(2)}|}$$

$$t'' = \frac{x_{1(1)} + x_{2(2)} - x_{2(1)} - x_{2(2)}}{|x_{1(1)} - x_{1(2)} - x_{2(1)} + x_{2(2)}|}$$

provided a t-quantity for testing $m_1=m_2$ with one degree of freedom. Both Fisher and Yates have objected to the element of choice in t' and t''. I agree this is a weakness, but my purpose was not to put forward the best test, but a valid test serving to refute the Behrens-Fisher solutions (which takes in effect the less divergent of t' and t'' as a t-quantity with one degree of freedom). The test I actually proposed and used amounted in this case to using an estimate of $\sigma_1^2+\sigma_2^2$ by taking the mean square of the denominators of t' and t'' and assigning it an unknown number of degrees of freedom between 1 and 2, and hence certainly as great as 1. (If $\sigma_1^2=\sigma_2^2$, it would have 2 d.f., whereas if either σ_1^2 or σ_2^2 were zero it would have only 1.) This proposed test, being both valid and superior to the use of t' or t'' emphasized even more, in my opinion, the 'inefficiency' of the Behrens-Fisher solution.

Yates (in the paper cited) discusses the case

$$x_{1(1)} = 10{\cdot}1, \qquad x_{1(2)} = 19{\cdot}7, \qquad x_{2(1)} = 16{\cdot}5, \qquad x_{2(2)} = 26{\cdot}2.$$

This example gives

$$|t'| = 0{\cdot}67, \qquad |t''| = 129, \qquad |T| = 0{\cdot}95.$$

I do not know whether the use of T provides the optimum test in some sense, but I certainly believe it superior to the Behrens-Fisher test. For such a small sample (really too small for *statistical* inferences) it seems difficult to pin down further the uncertainty in the number of degrees of freedom for T by making use of the observed variance ratio s_1^2/s_2^2, though for larger samples Welch (1947) has shown how this may be done.

Incidentally, it has been shown, for example, by Jeffreys, that the Behrens-Fisher solution is the Bayes solution for uniform and independent prior distributions for m_1, m_2, log σ_1 and log σ_2. This is hardly a result to be used in support of *fiducial* probability; but in any case I would not regard such prior distributions as very sensible in this problem, as they are incompatible with any finite and non-zero observed value for s_1^2/s_2^2 to be used in the solution (cf. James *loc. cit.* p. 80).

Fisher had in his 1935 paper generalized rather tentatively to the concept of simultaneous fiducial distributions, appearing to argue largely by analogy from the case of a single normal population with unknown mean and variance, a problem where no difficulty arises. In my own paper I noted that the single population case could be extended to any joint problem of location and scaling, in the sense that valid and consistent classes of confidence intervals for either parameter, when the other was known or unknown, could be generated from the two-parameter fiducial distribution. This result was independently noted by Pitman in 1938. When valid confidence intervals do not exist, I do not consider that a case for a separate class of fiducial intervals, derived independently either of the theory of confidence intervals or of inverse probability, has been established. A good deal more has been said or written since these early papers, but I cannot see that it has taken us much further forward. This applies, for example, to the discussion on fiducial probability at the I.S.I. meeting in 1953 at Ottawa,* where indeed support for my doubts on the value of fiducial inference as such will be found, for example from Lindley and Pitman, though admittedly not in every instance from the same side of the Bayesian fence. This view would not of course be supported by others, such as Barnard and Birnbaum, the latter writer's notion of 'intrinsic confidence intervals' being apparently close to Fisher's fiducial intervals. But I do not feel disposed to accept further concepts unless they have some demonstrable practical value. I can see some pros and cons in the case of the Bayesian, non-Bayesian, controversy, depending on one's attitude to the relative status of statistical probabilities and probabilities as degrees of belief. Both these distinct probability concepts have, whatever our personal predilections, considerable acceptance and recognition, both from statisticians and philosophers; yet more probability concepts, with no very clear interpretation, are hardly something we can welcome.

4. BAYESIAN INFERENCE

Having digressed for too long on fiducial inference, I hope I shall not do the same with Bayesian inference. However, some comment is unavoidable, as, in spite of Fisher's onslaught on inverse probability during his lifetime, an attack with which, with some reservations, I concur, the Bayesian approach is still very much with us. Indeed, from the efforts of Jeffreys, Savage, Good, Lindley and others, its status has been growing of late. This has partly been due to the development of decision theory and the introduction of loss functions. Nevertheless, as I see it, there are one or two fundamental issues in the use of Bayesian methods in statistical inference that are too often ignored or played down by one side or the other.

I have reminded you that there are two distinct probability concepts. One, the statistical, is narrower in scope but more precise within its proper context than the other, the notion of degree of belief.† The subject of statistics has not only an almost universal range of application, but its meaning has also tended to broaden from the orthodox one of being concerned with population or group

* See Linnik *et al.* 1964.

† In answer to a query by F. J. Anscombe after my lecture on what I mean by this remark, perhaps I could refer to my note (1936b), the last sentence of which concludes: '(Statistical) probabilities may be said to exist objectively in the usual and necessary sense—that they are theoretically measurable, and sufficiently well substantiated by experiment.'

phenomena. I think this last trend has been rather confusing. Statistical phenomena have their own properties associated with laws of large numbers and ergodic theory that are quite separate logically from our overriding and constant necessity to be making decisions and inductions in any field, whether it is statistical or not. If a comprehensive and unique theory of induction were acceptable for all phenomena, as some Bayesians would claim, then statistical phenomena would naturally be dealt with in the same way. There is plenty of evidence (which as good statisticians we must not suppress) that this ideal situation does not exist. As this is so, it is open to the statistician to analyse his material in a manner which he thinks useful and explicable both to himself and to others; and to this end he has developed various techniques and methods especially designed to reduce and simplify statistical data. Further than this I do not think we can go. It is not surprising that the statistician in the search for objective results has often formulated them in terms of statistical probabilities, for example, statistical confidence in conclusions—the so-called 'behaviouristic' approach, as Neyman has put it. Certainly, this type of procedure has its limitations and dangers, but there is a fundamental impasse here that should be clearly stated—I do not think, for example, that Fisher was ready to admit it, although, somewhat ironically, the small-sample theory which Fisher helped so much to develop rather brought it into the open. The statistical approach cannot by its nature deal with the unique sample—it must contemplate statistical variation which often from the Bayesian angle is irrelevant and a source of inefficiency. The statistician can attempt to reduce the effect of irrelevant variation by conditional inferences and the like, but he cannot eliminate all questions of sampling variability or he has no probability distribution to work with at all. But the Bayesian cannot deal with the unique sample either except by moving into a different field of discourse, the quantitative aspects of which are debatable. There are various distinguishable schools. The one including Good and Savage assign personal degrees of belief and utilities to relevant propositions. This seems to me of possible value to an individual, for example, in business, assessing his own different courses of action. I am not convinced, however, that it can claim scientific validity unless the degrees of belief can be generally accepted and hence idealized. This is the standpoint of Jeffreys and possibly of Lindley. It is, for example, not unreasonable to assign a uniform conventional prior distribution to denote prior ignorance in the case of a finite and discrete set of alternatives. This is in line with information theory in the communication theory sense (i.e. Shannon's concept, not Fisher's). This is, I believe, as far as the logician Rudolf Carnap got; and there are considerable difficulties with any more complicated but equally common prior distribution problems. It seems in any case meaningless to me to claim quantitative induction in a unique situation. Some common features with other similar problems are recognized. Indeed, I think Reichenbach went so far as to claim that no inductions could be formalized except on a frequency basis. Concentration on the unique sample can be carried too far. Yet once some behaviouristic or frequency justification of inductions is sought, the Bayesian has an additional problem to contend with. It is then necessary to distinguish between his assigned prior distribution and the true prior distribution, which certainly exists in some cases in a frequency sense. Of course, in some sufficiently well-defined statistical situations a whole process can be studied as one problem, and the so-called empirical Bayes' procedures of Robbins seem to fit in here. But in a typical problem, say in decision theory, whereas most statisticians would grant that with the right prior distribution (in the frequency sense) and loss functions, the Bayes approach (including loss functions) would be the optimum one, and therefore certainly worth formulating, it is not always made clear what happens if, as is more likely, the wrong quantities are used.

The Behrens-Fisher test seems to afford an example where the use of a criticizable prior distribution leads to a solution inferior to other possible tests. Of course decision theorists have been aware of this problem ever since Abraham Wald developed decision theory and suggested the use of the minimax principle; but neither he nor later workers were very satisfied with this particular suggestion.

At present there is no final reconciliation of the Bayesian and non-Bayesian approaches; but at least the situation is hopeful in that there seem signs of less dogmatism, and more appreciation and tolerance of these respective viewpoints. I think even the Bayesians have appreciated some of Fisher's concepts such as sufficiency, likelihood and information, although they may be less relevant to their own approach (cf., however, Pratt, 1964) and although, like some of the rest of us, they may query the point of any of them, if put forward as concepts in their own right without the need for some acceptable probability interpretation. Some years ago (in my Inaugural Lecture at University College) I expressed the view that a final resolution of the inverse probability controversy must await a resolution of the question how much objectivity to assign to science and scientific theories. We may now admit Karl Pearson's thesis that scientific laws are man-made, but this does not prevent our proceeding on the lines that there is an objective world of phenomena to be described. A statistician who sets up a statistical model to be tested or whose parameters are to be estimated is proceeding along a well-known scientific path described by R. B. Braithwaite as following the hypothetico-deductive method. My own inclination is to follow this path. My impression is that it was Fisher's inclination also, at least in much that he did; but his claims for some kind of absolute validity and objectivity for some of his concepts such as fiducial inference seem to me rather to have fogged the issue.

There is some interest in one of the most recent (1962) notes of Fisher's in which he tried to show that inverse probability could in suitable cases be formulated in unequivocal terms. The type of example he took was one where particles are being emitted randomly at an unknown rate, and the investigation was arranged in two parts, the first part consisting of the random time to the first emission, and the second part containing some further statistical information on the same unknown parameter. Fisher formulated his solution in the terms of inverse probability, but as far as I can see he was merely combining two independent pieces of statistical information and no new situation has been engineered. His discussion merely seems to me to emphasize a failure in his later years to make it clear to others what his probability inferences for an unknown parameter meant. As he had by this time rejected the frequency interpretation the only other recognized interpretation is a degree of belief, but if so interpreted his formulation has to compete with more orthodox Bayesian approaches. In spite of some obvious advantages if an acceptable system of inference in terms of the statistical data alone could be formulated, I can see no evidence that Fisher's later point of view has really been helping the development of practical statistical methodology, and this to me at least is perhaps the most important criterion by which to judge.

I am afraid I have after all spent rather longer on these controversial issues than is perhaps justified, bearing in mind my expectation that they will remain with us for some time—certainly I have been longer than I originally intended in a discussion primarily on statistical methodology. To recapitulate, it has been suggested that the basic issue is on Bayesian versus non-Bayesian methods, and that Fisherian variants on the non-Bayesian approach should not obscure this. It has, however, also been pointed out that there are at least three Bayesian approaches, the individualistic or personal approach of Savage, the epistemological approach of Jeffreys, and the prior frequency approach often attributed to Karl Pearson. In earlier writings on this subject I have where clarity demanded it used different probability notation for different concepts, and I at least would find writings by others clearer if the same practice were followed, especially now that the different probability concepts have multiplied.

DIFFERENT PROBABILITY CONCEPTS WITH POSSIBLE NOTATIONS*

Concept	Bartlett (extended)	Carnap (extended)
1. Rational degree of belief (epistemological approach e.g. Jeffreys)	$\underline{P}$	$\underline{p_1}$
2. Statistical probability or chance (assumed)	$\underline{p}$	$\underline{p_2}$
3. Degree of belief (personal or individualistic approach e.g. Savage)	iP	ip_1
4. Estimated or guessed, e.g. prior, probability (frequency approach)	ep	ep_2
5. Fiducial probability (where equivalent confidence probability)†	$\underline{fp}$	fp_2
6. Fiducial probability (Fisher)	fP	fp_1

* Underlining denotes previous usage.

† In answer to a referee, confidence probabilities that are not fiducial probabilities might where necessary be denoted by cp (or cp_2).

It will be a relief now to turn to less debatable issues. Could I just note first, however, how our basic approach may unconsciously affect our attitude to particular branches of statistical methodology. There has, for example, been invaluable work done in recent years on 'non-parametric methods.' These methods are often much more 'robust' against the correctness of background assumptions about the model used, and therefore have an important rôle to play in modern statistical methodology; but I would be reluctant to see the statistician relinquish his responsibility to set up explicit statistical models which represent, as far as he can manage, the situation he is investigating and analysing. Indeed, with the coming, say, of operational research, or of mathematical biology, the use of models has been spreading, and their use by the statistician becoming more diverse and intricate.

5. THE DEVELOPMENT OF EXPERIMENTAL DESIGN AND ANALYSIS OF VARIANCE

Fisher's quest for precise methods of statistical inference led him to make tremendous advances in the technique and analysis of statistical experiments, his contributions in this general field being as important practically as anything else he did in statistics. From his day-to-day contact with agricultural experiments at Rothamsted, Fisher came to realise the essentials of good experimental design. He perceived the simultaneous simplicity and efficiency of balanced and orthogonal experimental designs. If any statistical assessment of error was to be possible, replication was of course necessary, and equal numbers of plots per treatment optimised the design and greatly simplified the 'analysis of variance' technique that was developed to go with it.

It was at this point that Fisher introduced one vital principle. When statistical data are collected as natural observations, the most sensible assumptions about the relevant statistical model have to be inserted. In controlled experimentation, however, randomness could be introduced deliberately into the design, so that any systematic variability other than due to the imposed treatments could be eliminated. This has been an invaluable device in practical experiments and sampling surveys of all kinds. Incidentally, it is a device that has been a source of some logical difficulty to the orthodox Bayesian (see Savage *et al.*, 1962, especially pp. 87–91).

The second principle Fisher introduced naturally went with the first. With the statistical analysis geared to the design, all variability not ascribed to the influence of the treatments did not have to inflate the random error. With equal numbers of replications for the treatments each replication could be contained in a distinct block, and only variability among plots in the same block were a source of error—that between blocks could be removed. This principle, like the first, is also of course of extreme importance in the design of sampling surveys.

The third principle Fisher introduced was in connection with treatment combinations of more basic factors, such as the testing of combinations of the three primary fertilizer ingredients, nitrogen, phosphate and potash, in agricultural trials. Fisher emphasized the gain by testing all combinations and breaking down the analysis into the separate degrees of freedom for main effects, first-order interactions and so on. This technique of factorial experimentation cut right across the current practice which Fisher criticized in his book on The Design of Experiments (1935b; p. 96) referring to the 'excessive stress laid on the importance of varying the essential conditions *only one at a time*.' He showed how the individual factors could be numerically assessed in the *absence of interaction* as if the other factors were absent, with, moreover, a wider inductive basis for the assessment. Moreover, if interaction were present, information was obtained on its effect that could not have been obtained by separate testing of the individual factors.

It will be noticed that the value of these methods was not dependent on the statistical analysis, but the simplicity and clarity of the relevant analysis, which Fisher emphasized was dictated by the design, greatly contributed to their rapid world-wide popularity. The analysis was basically classical least-squares analysis, but the orthogonality of the design rendered the estimation problem trivial. The associated estimation of error was systematized by the technique of 'analysis of variance,' perhaps a slightly unfortunate title as the analysis mainly consists of a numerical breakdown of the total sum of squares of the observations into its relevant additive parts. Once the technique had evolved (which did not happen overnight) and the appropriate significance tests were made available from statistical tables, based on Fisher's derivation and tabulation of the variance-ratio distribution, more complicated least-squares problems, such as non-orthogonal designs or multiple regression analysis, could also be dealt with.

It might be remarked that all this technical advance in experimental design had its dangers. The design and analysis were logically more tied to the null hypothesis of no treatment effects than to any alternative, and this gave somewhat undue importance to the rôle of the significance test. The *validity* of the methods for small samples was sometimes confused with their *sensitivity*. The complications that might ensue once the null hypothesis was false were not always followed through. The additive set of interactions in factorial experiments was always logically correct, but less practically relevant in some contexts than in others. Such limitations are, however, not dissimilar logically from those in other statistical fields of analysis. It is a tremendous practical gain to have simple, efficient routine methods of analysis, always provided that they are not elevated to a blind ritual.

6. MULTIVARIATE ANALYSIS

One of the useful extensions of analysis of variance technique for the analysis of experiments was the so-called analysis of covariance technique for adjusting final observations by initial ones made before treatments have been imposed. This technique is of one personal interest to me as being, back in 1933, the first instance I had that Fisher, like the rest of us, could err, in this case at least by implication. At that time he had merely given the adjustment without stressing that the ensuing non-orthogonality necessitated further analysis if an exact test of significance of treatment effects was required.

The simultaneous analysis of variance and covariance is of course something different, being an example of multivariate analysis as more usually defined. Fisher shares primarily with Hotelling in the United States and Mahalanobis in India the distinction of developing this technique, which has with the greater availability of computers been growing of late in importance. I have occasionally noticed a tendency to denigrate Fisher's astonishing power of geometrical reasoning, which I mentioned earlier, and which was to assist him in obtaining so many solutions of sampling distribution problems. It was the same approach that enabled Wishart to obtain with Fisher's guidance the so-called Wishart distribution. It was the same approach too that enabled Fisher in 1928 to obtain the general distribution of the multiple correlation coefficient, a derivation that few people understood. I remember when lecturing in Cambridge before the war preferring Wilks' analytical derivation for this very reason. When I returned after the war, J. O. Irwin happened to raise the matter again in conversation, and I returned to Fisher's orginal derivation. This time I was delighted that I could follow it, and at once realized that his argument could be extended in principle to the general distribution of canonical correlations, a problem with which I was in consequence able to make some headway.

It was always something of a surprise to me that Fisher did not himself continue to make full use of geometrical argument in establishing the sampling theory for multivariate analysis. The sampling theory of his linear discriminant function was implicit in Hotelling's earlier work on the multivariate extension of the *t*-test, but the effect of eliminating a hypothetical discriminant function (or even a hypothetical first canonical variate in the more complex sampling problems of canonical correlation analysis) could also be studied by means of the reciprocal sampling relations between two sets of variables. Fisher, by attempting to proceed rather formally by pseudo-analysis of variance technique, was led at times into definite errors.*

7. TIME SERIES AND STOCHASTIC PROCESSES

One man, even of Fisher's calibre, cannot of course maintain the same level in all relevant areas. In retrospect, one can attempt to see Fisher's work against the general scientific background; and my own impression is of one rather serious omission in its coverage. His comparative neglect of the important Continental work on the foundations of mathematical probability I have already suggested was on the whole not practically very important. This neglect, however, extended to developments in the theory of random or stochastic processes, even when such developments were not of the more fundamental type associated with such contemporaries as Kolmogorov, Khintchine and Slutsky in the U.S.S.R., or Wiener, Feller and Doob in the U.S.A., but were associated with the more immediate problems of statistical theory and analysis such as the work in his own country of G. U. Yule on time series and A. G. McKendrick on stochastic models in biology.

This was particularly surprising in view of Fisher's active interest in stochastic processes in genetics. Thus Fisher's papers on the statistical analysis of data recorded in time included some interesting work on orthogonal polynomial fitting, but a rather rigid adherence to classical least-squares procedures. His 1929 paper on the 'Studentisation' of the classical significance test of a strict harmonic component (of unknown periodicity) came later than Yule's famous 1927 paper on autoregressive models for stationary time-series, but no discussion of the repercussion of Yule's work on the whole subject of periodicity in time-series, or indeed any reference at all to it, was made. It should not be necessary to remind you of the tremendous progress made with time-series analysis since then, progress which is making this subject one of the most rapidly developing branches of statistical methodology at the present time.

---

* *Cf.* for example, Fisher's discussion in *Statistical Methods for Research Worker's*, 10th Ed. (1946), Example 46.2 and my own (Bartlett, 1951–2, §4).

The work of McKendrick was less directly connected with statistical methodology, but fundamental in the development of stochastic models in biology and medicine. Moreover, it had indirect relevance to the study of 'contagious-type' distributions by Greenwood, Yule, Polya and Neyman. Here Fisher was on common ground with these other workers in the overlooking of McKendrick's work, which included, for example, the derivation in 1914 of the negative binomial distribution as a contagious-type distribution.

The problems of statistical analysis and statistical methodology arising in the general area of stochastic processes are very important and complex, as I have indicated already elsewhere (1959). They are certainly not confined to the problems of time-series analysis in the usual sense, and are rapidly growing as the use of stochastic process models develops in biology, economics, industry, medicine and psychology.

8. CONCLUDING REMARKS

It would be foolish of me to pretend I could adequately assess all the contributions to statistical methodology made during the last fifty years. I have not even mentioned yet the brilliant work done by Abraham Wald during the last war on the principles of sequential sampling. There are plenty of important developments on the more strictly practical level, such as the great development in the principles and practice of sampling surveys, or the development of cohort analysis in demographic studies. I am sure I shall be accused, especially on this side of the Atlantic, of underemphasizng the rôle of decision theory in modern statistics, my rather cursory references to it being associated with my view that (i) it is not properly classifiable as statistics, as I understand this term (I feel tempted to say it is more a way of life!), (ii) it involves concepts such as loss functions and prior probabilities which I find quantitatively to be of dubious practical value in the study of statistical phenomena, at least in the scientific field.

Obviously I have, while ranging to some extent outside Fisher's own activities, made these the starting point of this survey. With regard to his own work, my somewhat protracted preoccupation with the more debatable issues will be disliked by his more dedicated admirers. The trouble with great men, especially those with temperaments of comparable stature, is that they are liable to excite either allegiance or rebellion. This does not facilitate an objective judgment. However, let me recall Professor M. Fréchet's words: (1963, p. 169):

"*Les statisticiens du monde entier savent quelle dette ils doivent à l'école statistique brittanique, et, en particulier, aux deux grands savants qui ont, l'un créé, l'autre transformé la statistique mathematique, Karl Pearson et Sir Ronald Fisher.*"

Fisher would no doubt have thought Fréchet only half-right; but for ourselves, we do not have to accept Fisher's complete infallibility in order to recognize his greatness as a scientist, and like Fréchet, acknowledge the permanent debt which we all as statisticians owe him.

REFERENCES

Bartlett, M. S. (1936a) "The information available in small samples." *Proc. Camb. Phil. Soc.* **32**, 560–6.

——— (1936b) Statistical probability. *J. Amer. Stat. Ass.* **31**, 553–5.

——— (1951–2) "The goodness of fit of a hypothetical discriminant function in the case of several groups." *Ann. Eugen.* **16**, 199–214.

——— (1959) The impact of stochastic process theory on statistics (published in *Probability and Statistics*, Stockholm, pp. 39–49).

——— (1961) "Probability, statistics and time." (Inaugural lecture, University College, London.)

Birnbaum, A. (1962) "On the foundations of statistical inference." *J. Amer. Stat. Ass.* **57**, 269–326.

Eddington, A. S. (1935) *New Pathways in Science.* (University Press, Cambridge.)

Fisher, R. A. (1912) "On an absolute criterion for fitting frequency curves." *Messenger of Mathematics.*

——— (1915) "Frequency distribution of the values of the correlation coefficient in samples from an indefinitely large population." *Biometrika* **10**, 507–21.

——— (1925) "Theory of statistical estimation." *Proc. Camb. Phil. Soc.* **22**, 700–25.

——— (1928) "The general sampling distribution of the multiple correlation coefficient." *Proc. Roy. Soc.* A **121**, 654–73.

——— (1929) "Tests of significance in harmonic analysis." *Proc. Roy. Soc. A.* **125**, 54–9.

——— (1930a) *The Genetical Theory of Natural Selection.* (University Press, Oxford.)

——— (1930b) "Inverse probability." *Proc. Camb. Phil. Soc.* **26**, 528–35.

——— (1935a) "The fiducial argument in statistical inference." *Ann. Eugen.* **9**, 174–80.

——— (1935b) *The Design of Experiments.* (Oliver and Boyd, Edinburgh.)

——— (1946). *Statistical Methods for Research Workers* (10th Ed. Oliver & Boyd, Edinborough.)

——— (1962) "Some examples of Bayes' method of the experimental determination of probabilities *a priori*." *J. R. Statist. Soc.* B. **24**, 118–24.

Fréchet, M. (1963) See obituary of Sir Ronald Aylmer Fisher, 1890–1962. *J. R. Statist. Soc.* A. **126**, 159–78.

James, G. S. (1959) "The Behrens-Fisher distribution and weighted means." *J. R. Statist. Soc.* B. **21**, 73–90.

Kolmogorov, A. N. (1933) *Grundbegriffe der Wahrscheinlichkeitrechnung.* (Springer, Berlin).

Linnik, Yu. V. and other contributors. (1963) "Fiducial probability." *Bull. I.S.I.* **40** (vol. 2) 833–939.

McKendrick, A. G. (1914) "Studies on the theory of continuous probabilities with special reference to its bearing on natural phenomena of a progressive nature." *Proc. Lond. Math. Soc.* (2) **13**, 401–16.

——— (1926) "Applications of mathematics to medical problems." *Proc. Edin. Math. Soc.* **44**, 98–130.

Neyman, J. and Scott, E. L. (1948). "Consistent estimates based on partially consistent observations." *Econometrica.* **16**, 1–32.

Pitman, E. J. (1938) "The estimation of the location and scale parameters of any given form." *Biometrika,* **30**, 391–421.

Pratt, J. W. (1965) "Bayesian interpretation of standard inference statements." *J. R. Statist. Soc.* B, **27**.

Savage, L. J. and other contributors. (1962) *The foundations of statistical inference.* (Methuen, London.)

Student (1908a) "The probable error of a mean." *Biometrika,* **6**, 1–25.

——— (1908b) "Probable error of a correlation coefficient." *Biometrika,* **6**, 302–10.

Welch, B. L. (1947) "The generalization of 'Student's' problem when several different population variances are involved." *Biometrika,* **34**, 35–8.

Yates, F. (1964) "Fiducial probability, recognisable sub-sets and Behrens' test." *Biometrics,* **20**, 343–60.

Yule, G. U. (1927) "On the method of investigating periodicities in disturbed series, with special reference to Wolfers's sunspot numbers." *Phil. Trans.* A. **226**, 267–98.

The paradox of probability in physics

Classical paradoxes

Probability is a controversial subject whether or not we consider it in relation to fundamental issues in physics, but most of the general controversy centres on its alternative interpretations in (i) statistical, or (ii) subjective, terms; and this problem is not in my opinion immediately relevant to the present discussion, for which I shall always use the first meaning. Moreover, at a meeting in 1956 of the Philosophy of Science Group of the British Society for the History of Science, I discussed (see Bartlett, 1962) the three classical paradoxes – Loschmidt's reversibility paradox, Zermelo's recurrence paradox and the paradox of Maxwell's demon – therefore I do not want to consider these in detail now. They all relate to the Second Law of Thermodynamics. Loschmidt's and Zermelo's were formulated in terms of deterministic laws, but they are not removed either by the introduction of quantum mechanics or by statistics. In fact, while Loschmidt's is partly resolved by the distinction between conditional and absolute probabilities, both paradoxes emphasize the limitations of the Second Law, which is not infallible for all sizes of physical system. Maxwell's paradox is, I believe, even more subtle, for Szilard showed that it could only be resolved by ascribing to observations an inevitable interference with the system leading to entropy increase at least as great as the reduction intended (cf. Brillouin, 1962. §13.6).

Probability and quantum mechanics

What I would like to do here is to concentrate on a comparatively more recent problem, the place of probability in quantum mechanics. Eve since my undergraduate days at Cambridge I have found the orthodox presentation unsatisfying, not just because the statistical basis is different but because it did not seem properly introduced or explained. Historically it was first put forward by Born as an afterthought, and has never to my mind been adequately clarified.

I should interpose at this point (in this written version) to say that I have been referred to Professor Mackey's elegant little book on the *Mathematical Foundations of Quantum Mechanics*, but I did not find that this removed my difficulties; indeed, it seemed to underline them by concentrating the crucial postulate into one axiom (No. VII) which the author himself admitted in his book to be both sophisticated and *ad hoc*. Still later, I had the advantage of further conversation with Professor Mackey, who recalled such papers as that by Varadarajan (1962), who has discussed the theoretical generalization of classical probability theory to include the simultaneous incompatibility of measurement familiar in quantum-theoretical formulations.

However, such theoretical developments (cf. also Landé, 1960), while they help to give a firmer basis to the orthodox structure of quantum mechanics, are still for me too abstract and formal and do not appear to rule out the possibility of further physical insight coming from other directions. Several illustrious physicists, including Einstein and de Broglie, have expressed doubts about the orthodox theory, though these two particular pioneers in the development of quantum theory appeared to favour a return to determinism, a step which would seem to me to be retrograde.

Perhaps here I could refer to one barrier to clarity which I have found remarkably inconsistent with the tremendous emphasis put by modern physicists on *operational* theories. It is pointless to argue overmuch about *unobservable* features of any approach. Two or more approaches which are observationally equivalent in all respects may be regarded as of equal merit, unless one of them gives either (i) greater technical facility, or (ii) greater insight. It is well-known that Schrödinger's wave-mechanics and Heisenberg's operational calculus were two such equivalent approaches, but the possibility of others is illustrated by the later introduction by Feynman of his path integral technique. The value of a more direct introduction of statistical concepts into any new formulation of quantum theory would lie in the possible avenues it might open up into new terrain, though orthodox quantum theorists can reasonably defer their own judgment until any such new approach has been successfully developed. At present such an approach appears still lacking, though I believe some tentative discussions are significant, intriguing, and worth examining further.

The obstacle I have referred to creates a difficulty for the comparative layman like myself, for it is not always easy to find out from textbooks what is observed and hence what has to be accounted for. The Uncertainty Principle in particular has even been held up as an article of faith which it would be heresy to query or discuss. On the question of quantization, I once read carefully through a standard introductory textbook on quantum mechanics, delving through the details of the discrete energy levels of the harmonic oscillator, and thence on to the hydrogen atom, searching for some discussion of observational checks. I could not find any until I reached an Appendix, which presented a rather belated discussion of the interaction of the hydrogen atom with an external electromagnetic field, the consequent spectral lines (no longer discrete!) being the first direct link with observation. Of course I have no objection to directly unobservable concepts such as the wave function, but it is crucial to discriminate clearly between what is observed and what is purely conceptual. The features of quantum theory which any new approach must certainly cope with include (a) the Uncertainty Principle (b) quantization (c) the Superposition Principle. The first two have often been mentioned as though they were the most paradoxical, but it is the third which I regard as being the most difficult to reconcile with statistical ideas. Let us consider them in turn.

The Uncertainty Principle. In spite of the 'heresy' of considering simultaneously conjugate observables such as position and momentum, to which I have referred, there seems nothing to stop us considering their measurement consecutively, provided we recognize that the earlier position measurement might be upset by a subsequent momentum measurement. This permits a cautious use of phase-space distributions in both variables, and these were in fact already introduced by Wigner in 1932 (cf. Moyal, 1949). Such a use led to the discovery that the free particle and the harmonic oscillator may be treated on classical deterministic lines, provided the correct initial phase-space distribution is inserted. This feature has some technical advantages which permitted Moyal and myself (1949) to give a general exact solution for the transition probabilities of the harmonic oscillator with a perturbing potential field, making use of a phase-space expansion in terms of the familiar eigenvalues $1/2h\nu$, $3/2h\nu$, ...

Such properties for the free particle and the oscillator seemed, however, somewhat fortuitous. They did not dispel the fundamental place of the Uncertainty Principle in quantum mechanics. If we are to 'explain' this principle, we might perhaps look for analogous principles elsewhere which are statistically more transparent. When we do, we notice a close analogue in the statistical theory of time-series, where there is an intrinsic uncertainty principle well-known to communication engineers connecting time and frequency (the reciprocal of periodicity). The transference from position-momentum to time-energy is immediate on relativistic symmetry (though this aspect of the Uncertainty Principle is often omitted, for example when discrete energy levels are under discussion), so that we have an encouraging link which would be even stronger if the quantum theory relation $E = h\nu$ connecting energy and frequency could be clarified further. The immediate point to notice is that uncertainty principles exist in other areas which are natural statistical principles, and do not have to be postulated as *a priori* axioms.

Quantization. The occurrence of discrete energy levels also seems to have some natural aspects. The existence of discrete eigenvalues for *bound* physical systems such as vibrating strings is a pointer to analogues elsewhere. Stochastic process solutions for such systems would be expected to be developed in terms of discrete eigenvalue expansions; and other known stochastic process solutions for decaying open systems, involving exponentially-decreasing time factors, have their analogues in quantum-theoretical problems of a similar nature.

In the case of the harmonic oscillator it has been suggested (Marshall, 1963) that the ground state can be thought of in terms of the equilibrium between the oscillator and residual radiation; the question that seems to arise here is whether the higher energy levels, which at least exist as further terms in a general distributional expansion for the

oscillator, exist separately, either for the oscillator or for any other
quantum system. The importance of knowing just what is observable
for such systems is apparent.

The Superposition Principle. This principle, which refers to the
addition not of probabilities but of wave-functions is, as already
remarked, perhaps the most baffling from the statistical standpoint. To
take specifically the case of electron interference, we may consider a
stream of electrons passing through two neighbouring slits and
producing an interference pattern on a photographic plate. This pattern
is obliterated if we uncover one slit at a time, but not if, in the presence
of both slits, we reduce the intensity of the stream so much that only
one electron can be hitting the plate at any instant, and we allow the
interference pattern to build up gradually from these individua
impacts. I will return to this problem after discussing some recen
theoretical investigations.

Recent speculative investigations.

One feature of many of the attempts in later years to find a more direc
statistical approach to quantum mechanics is to postulate a basi
Brownian motion affecting all phenomena. These attempts often hav
some attractive aspects, but appear to fail to account for all the relevan
observational phenomena, and indeed are rarely taken far enough to b
checked in this way. Moreover some theoretical care is necessary whe
dealing with Brownian motion, especially in relation to the velocity v o
a particle. If Brownian motion is postulated in relation to increments i
the position coordinates of the particle, its velocity is undefined; if i
relation to the velocity of the particle, then v exists, but its averag
square is not equal to its squared average. A paper by Kershaw (1964
appears to me to contain theoretical inconsistencies, or at least lacunae
of this sort; that by Nelson (1966) appears more consistent but require
a peculiar and not obviously admissible assumption of zero viscosity.

If we return for a moment to Feynman's path integral formul
tion, this, in so far as it has been made rigorous, appears equivale
to orthodox quantum theory, but has no immediately obvio

justification on any stochastic process approach. Gilson (see, for example, 1968) has made some detailed investigations of this formulation, and has established some useful restrictions on the nature of any compatible stochastic theory. In particular, he has derived the Wigner-Moyal phase-space distribution directly from the path integral; and has also given a more general formulation which includes quantum theory as a limiting case. Nevertheless, while Gilson's investigations are not complete, they do not seem to suggest any very great departure from orthodox theory. This might of course be an advantage, but more radical departures from orthodoxy also seem to merit study.

An intriguing paragraph in the book by Feynman and Hibbs indicates an alternative derivation of the quantum-mechanical motion of a particle. It could be maintained that Feynman's path integral formulation envisages already a random basis for the motion, but this is hardly the same as some specific stochastic process derivation, for the Feynman path integral is still that for the (complex-valued) wave function, and not for a probability as such. However, within this theoretical framework, the alternative approach is based on a particle with random transitions from velocity $+c$ to $-c$. This approach even appears consistent with the relativistic modification of orthodox theory, and with remarks made by Dirac (1935) on the possible values for the velocity of a particle. I am indebted to H. A. C. Dobbs for this reference to Dirac, and for further discussion bearing on the compatibility of this viewpoint with the possibility of ascribing a non-zero mass m to the particle. Whether or not there is any inconsistency here, the *ad hoc* introduction of mass in the Schrödinger (or Dirac) equation seems an additional arbitrary feature which one day will be eliminated.

There was thus considerable interest for me in an investigation by Cane (1967), using a Brownian motion approach combined with the $\pm c$ constraint, because mass emerged as an intrinsic parameter of the entire system, not of the particle alone. Let me try to summarize some of Dr Cane's results (I hope fairly) and then comment on them.

A random walk motion is postulated on a discrete lattice, one-dimensional spatially to begin with, and in discrete time. The possible steps are ± 1 on the lattice, with probability p for $+1$, giving a mean velocity v forwards proportional to $2p - 1$. If the lattice interval is κ, and the time between lattice points is τ, then $\kappa = c\tau$, and $v = (2p - 1)c$.

The variance in position x after time t is

$$\sigma^2(x) = \sigma_0^2 + 4\kappa^2 tp(1-p)/\tau \sim \kappa ct$$

if $\sigma_0^2 = 0$, and $p \sim \frac{1}{2}$. More precisely, when $\sigma_0^2 = 0$,

$$\sigma^2(x) = \kappa ct(1 - v^2/c^2).$$

If the 'momentum of the particle is defined to be $p_x = m_0 x/t$, then its variance

$$\sigma^2(p_x) = m_0^2 \kappa c(1 - v^2/c^2)/t$$

so that approximately

$$\sigma^2(p_x)\sigma^2(x) \sim m_0^2 \kappa^2 c^2 = \hbar^2/4$$

if m_0 is identified as $\frac{1}{2}\hbar/(\kappa c)$. More precisely, we may write $p_x = mx/t$, where

$$m = m_0(1 - v^2/c^2)^{-\frac{1}{2}},$$

and obtain exactly

$$\sigma^2(p_x)\sigma^2(x) = \hbar^2/4.$$

The approximating diffusion equation when κ and τ are assumed small (but $\kappa^2\tau$ not small) is discussed*, and also the generalization to correlated random walks, when the probability of a positive step depends on the direction of the previous step. The equations are also extended to three dimensions, and various similarities with wave mechanics noted; in particular the Schrödinger equation for a particle moving subject to a central potential $V = -Ze^2/r$ is derived from a Brownian motion with constant mean drift towards the centre. Such similarities are provocative. However, an attempt is then made to explain interference by means of an input of particles of oscillating intensity $A(1 + \cos \omega t)$. Even after this *ad hoc* introduction of a periodic term, I fail to see how this *temporal* periodicity would produce an observable *spatial* interference pattern. It is shown by more or less

* In the Appendix, I show that these random walk results may be derived directly in continuous space and time, and thus have no dependence on any 'granular structure for space and time.

standard arguments that at transverse distances x on a screen some distance D from two slits, separated transversely by 2δ and a distance a from the source, that the intensity at x is

$$A\{2 + \cos\{\omega(t - [(x - \delta)^2 + D^2]^{1/2}/v)\} + \cos\{\omega(t - [(x + \delta)^2 + D^2]^{1/2}/v)\}\},$$

and this is constant when

$$x \sim D\lambda(2n + 1)/(4\delta),$$

fluctuating at intermediate values of x. It is claimed that the constant intensity provides no signal whereas the oscillating one does, but this explanation is unconvincing, as the integrated intensity over time will be the same at all points. Thus this mechanism could still not produce different effects at different values of x unless the temporal effect on, say, the photographic plate at the screen were a non-linear one. Not only have I been assured by physicists that this is not the case, but it would seem irreconcilable with the known invariance of the interference phenomenon to the instantaneous temporal intensity, which can be reduced indefinitely. The paradox of interference is that it arises from 'self-interference' when *both* slits are open, so that no explanations in terms of *particle* models, at least of the above kind, appear possible. (This comment is consistent with my remarks in my University College Inaugural Lecture [§III(2), at end], but not with the view of Landé (*op. cit.* Ch. III), whose axioms for a probability metric consistent with wave mechanics I find obscure. I am more inclined to favour approaches in terms of *events* as fundamental entities rather than *particles*, which Schrödinger once referred to as particular kinds of event sequences.)

Appendix

Consider random motion in one dimension, with transition probabilities αdt of reversing from $+c$ to $-c$ in velocity, and βdt of reversing from $-c$ to $+c$, in $t, t + dt$.

Then

$$M_t(\theta,\phi) \equiv E[e^{\theta X_t + \phi \dot{X}_t}]$$
$$= e^{\phi c}E_+[e^{\theta X_r}] + e^{-\phi c}E_-[e^{\theta X_r}]$$
$$= e^{\phi c}M_t^+(\theta) + e^{-\phi c}M_t^-(\theta).$$
$$M_t(\theta,0) = M_t^+(\theta) + M_t^-(\theta).$$

We have the forward equations

$$\begin{cases} \partial M_t^+/\partial t = \theta c M_t^+ + \beta M_t^- - \alpha M_t^+ \\ \partial M_t^-/\partial t = -\theta c M_t^- + \alpha M_t^+ - \beta M_t^-, \end{cases} \tag{3.1}$$

whence

$$M_t = (1 \quad 1)\begin{pmatrix} M_t^+ \\ M_t^- \end{pmatrix} = (1 \quad 1)\, e^{\underset{\sim}{A}t}\begin{pmatrix} M_0^+ \\ M_0^- \end{pmatrix}$$

where

$$\underset{\sim}{A} = \begin{pmatrix} \theta c - \alpha & \beta \\ \alpha & -\theta c - \beta \end{pmatrix}$$

For large t, the distribution of X_t has asymptotic cumulant function $\Lambda(\theta)t$, where $\Lambda(0) = 0$ and

$$(\theta c - \alpha - \Lambda)(-\theta c - \beta - \Lambda) - \alpha\beta = 0$$

i.e.

$$\Lambda = -\tfrac{1}{2}(\alpha+\beta) + [\theta^2c^2 + \theta c(\beta-\alpha) + \tfrac{1}{4}(\alpha+\beta)^2]^{1/2}$$
$$= \frac{\theta c(\beta-\alpha)}{\beta+\alpha} + \frac{\theta^2c^2}{\beta+\alpha}\left[1 - \frac{(\beta-\alpha)^2}{(\beta+\alpha)^2}\right] + 0(\theta^2), \tag{3.2}$$

which is equivalent to a normal asymptotic distribution with mea[n] $(\beta-\alpha)ct/\beta+\alpha$ and variance $8\alpha\beta c^2t/(\beta+\alpha)^3$.

This is compatible with Cane's results, with the identificatio[n] $2p - 1 = \epsilon = \beta - \alpha/\beta + \alpha$, $\kappa = 2c/\beta + \alpha$.

It should, however, be noticed that this *asymptotic* result [i] equivalent to the solution of the Brownian-motion type equation of the familiar form in the over-all density.

$$f_t(x) = f_t^+(x) + f_t^-(x)$$

viz

$$\frac{\partial f_t(x)}{\partial t} + m\frac{\partial f_t(x)}{\partial x} = \tfrac{1}{2}\sigma^2 \frac{\partial^2 f_t(x)}{\partial x^2} \tag{3.3}$$

where $m = E[\mathrm{d}X_t]/\mathrm{d}t$, $\sigma^2 = E[(\mathrm{d}X_t)^2]/\mathrm{d}t$, are the increase in mean and variance per unit time. More generally, when m and σ^2 conditional on $X_t = x$ are of the form $m(x)$, $\sigma^2(x)$, this equation is replaced by its generalisation with $m(x)$ and $\sigma^2(x)$ inside the differentiation signs. A stationary solution $f(x)$ of this equation may then exist, with

$$f(x) \propto \frac{1}{\sigma^2(x)} \exp \int^x \frac{2m(x)\mathrm{d}x}{\sigma^2(x)} \tag{3.4}$$

(see, for example, Bartlett, 1966, §3.5, equation (3.13)). No other stationary density solution of the equation exists, and the question arises whether more than one can exist for the exact equation, which in terms of $f_t^+(x)$ and $f_t^-(x)$ becomes (now for β and α functions of x)

$$\left.\begin{aligned} \partial f_t^+/\partial t + c\,\partial f_t^+/\partial x &= \beta(x)f_t^- - \alpha(x)f_t^+ \\ \partial f_t^-/\partial t - c\,\partial f_t^-/\partial x &= \alpha(x)f_t^+ - \beta(x)f_t^- \end{aligned}\right\} \tag{3.5}$$

By addition,

$$\partial f_t/\partial t + c\,\partial g_t/\partial x = 0, \tag{3.6}$$

where $g_t = f_t^+ - f_t^-$. For a stationary solution, $\partial g/\partial x = 0$, or $g = 0$, as no additive constant is possible for unlimited x. The equation for $f = f^+ + f^-$ then becomes

$$c\,\partial f/\partial x + [\alpha(x) - \beta(x)]f = 0,$$

or

$$f \propto \exp \int^x \frac{[\beta(x) - \alpha(x)]}{c}\,\mathrm{d}x, \tag{3.7}$$

a solution analogous to (3.4) above, with $\sigma^2(x)$ constant.

For example, if $\alpha(x) = \alpha x$, $\beta(x) = \beta x$ (with $\beta > \alpha, x > 0$) then

$$f \propto \exp\left[-\tfrac{1}{2}\frac{(\beta - \alpha)x^2}{c}\right] \tag{3.8}$$

This solution holds also for $x < 0$, if we reverse the sign of c for negative x, the equation for $\partial f/\partial x$ being then unaltered when we reverse also the sign of x, in the case $\alpha(x) = \alpha x$, $\beta(x) = \beta x$, $(x > 0)$.

If alternatively we consider the case of a circle, with $0 < y < 2\pi$, a solution

$$f^+ = f^- + C$$

if possible. However, the conditions

$$f^+(2\pi) = f^+(0), \qquad f^-(2\pi) = f^-(0),$$

now lead to the single solution

$$f^- = f^+ - C = \frac{C}{c}\,\xi(y)\left[\int_0^y \left\{\frac{\alpha(y)\mathrm{d}y}{\xi(y)} + \xi(2\pi)\int_0^{2\pi} \frac{\alpha(y)\mathrm{d}y}{\xi(y)[1 - \xi(2\pi)]}\right]\right. \tag{3.9}$$

where $\xi(y) = \exp[\int_0^y \left\{\beta(y) - \alpha(y\right\} \mathrm{d}y/c]$. For $\alpha(y)$, $\beta(y)$ of the fairly general form $\alpha\kappa(y)$, $\beta\kappa(y)$, this reduces to the *uniform* distribution $f(y) = 1/(2\pi)$, with

$$f^+(y) = \frac{\beta}{2(\beta - \alpha)\pi}, \qquad f^-(y) = \frac{\alpha}{2(\beta - \alpha)\pi}. \tag{3.10}$$

The important point is that the more precise formulation has not yielded more than one strictly stationary solution. The solution (3.8) is equivalent to the stationary solution (3.4) with $\sigma^2(x)$ constant, and $m(x) = -mx(x > 0)$.

This conclusion, which is consistent with the remarks on the harmonic oscillator in the main text, is emphasized because it appears to be at variance with the claims made by Nelson (1966), who suggests that the Schrödinger equation is entirely equivalent to Brownian motion of the type leading to (3.4) (with $\sigma^2(x) = \hbar/M$, where M is the mass). While both Nelson and Cane (*loc. cit.*) discuss the three-dimensional extension, this does not seem necessary for clarifying the points at issue, and it is simpler to restrict our attention to the one-dimensional case. Nelson defines a mean forward velocity synonymously with $m(x)$ in (3.4), and a mean backward velocity which he notes is expressible as

$$n(x) = m(x) - \sigma^2 \, \partial \log f_t / \partial x. \tag{3.11}$$

He claims that the 'state' of the particle must be described by its position X_t, its 'osmotic velocity' $\frac{1}{2}[m(x) - n(x)]$ and its 'current velocity' $\frac{1}{2}[m(x) + n(x)]$. This, however, appears to me to confuse the *state* of the particle for a Markov process, which is given by X_t for a process like (3.3) (or by X_t, U_t for a Markovian process in the velocity U_t), and the *specification* of the process, which in the case of normal diffusion for X_t is given by the specification of $m(x)$ and $\sigma^2(x)$. Thus while Nelson (cf. Cane, who writes $f_t = \sqrt{f}\psi$, so that ψ is real) writes

$$\psi(x,t) = \exp(R_t + iS_t), \tag{3.12}$$

where $R_t^2 = \log f_t$, and $\sigma^2 \partial S_t / \partial x = \frac{1}{2}[m(x) + n(x)]$, there is no explanation of why this is necessary.

In the case of the harmonic oscillator, the above discussion on Brownian motion may seem to contradict the conclusion of Moyal (1949) that its quantum-mechanical motion appears deterministic. This apparent inconsistency may presumably be resolved as follows.

Consider the deterministic motion

$$\begin{aligned} \mathrm{d}X_r &= U_r \, \mathrm{d}t \\ \mathrm{d}U_t &= -\omega^2 X_r \, \mathrm{d}t \end{aligned} \tag{3.13}$$

In this case the equation for

$$M_t(\theta,\phi) = E[\exp(i\theta X_r + i\phi U_t)]$$

(cf. Bartlett, 1966, §5.21) is

$$\frac{\partial M_t}{\partial t} = \left(\phi \frac{\partial}{\partial \theta} - \omega^2 \theta \frac{\partial}{\partial \phi}\right) M_t \tag{3.14}$$

with stationary solution satisfying

$$\left(\phi \frac{\partial}{\partial \theta} - \omega^2 \theta \frac{\partial}{\partial \phi}\right) K = 0 \tag{3.15}$$

where $K = \log M$. A solution of (3.15) is

$$K = A(\theta^2 + \omega^2 \phi^2), \tag{3.16}$$

a normal phase-space distribution which only corresponds to the quantum-mechanical ground state by appropriate choice of A. This choice is automatic via the Wigner-Moyal phase-space distribution, which starts from an initial distribution compatible with the Uncertainty Principle. From the present approach, the solution (3.16) might be related to the solution (3.8), which arises from deterministic motion between transitions, but has the advantage of leading naturally to a stationary distribution with no arbitrary constant. In fact, the Uncertainty Principle is satisfied with the identification $p_x = \pm Mc$, $\hbar/M = 2c\sqrt{[c/(\beta - \alpha)]}$.

The unsatisfactory character of any Brownian motion models of quantum mechanics not including the velocity (or momentum) explicitly in the state-specification may be clearly demonstrated in the even simpler example of one-dimensional free motion, for which the Schrödinger equation is

$$\frac{\partial \psi_t(x)}{\partial t} = \frac{\hbar i}{2m} \frac{\partial^2 \psi_t(x)}{\partial x^2} \tag{3.17}$$

Following Nelson's discussion (1967, p. 138) we write $\hbar/m = 1$ for convenience, and consider the initial conditions

$$\psi_t(x) = e^{-\frac{1}{2}x^2/a} \tag{3.18}$$

(so that $\sigma^2(X_0) = \frac{1}{2}a$). The relevant solution of (3.17) is

$$\psi_r(x) \propto (a + it)^{-\frac{1}{2}} \exp -\tfrac{1}{2}x^2/(a + it),$$

where

$$f_t(x) = \psi_r(x)\psi_r^*(x) \propto (a^2 + t^2)^{-\frac{1}{2}} \exp[-ax^2/(a^2 + t^2)]. \tag{3.19}$$

As with the harmonic oscillator, the solution (3.19) is consistent with deterministic free motion

$$T_t = U_0, X_t = X_0 + U_{0t},$$

provided $\sigma^2(X_0)\sigma^2(U_0) = \frac{1}{4}$, i.e. $\sigma^2(U_0) = 1/(2a)$.

Note that the phase-space distribution equation in x and u (as $m(u, x, t)$ and σ^2 for this equation are both zero) is simply

$$\partial g_t(x,u)/\partial t + \partial [u g_t(x,u)]/\partial x = 0, \tag{3.20}$$

with relevant solution

$$g_t(x,u) = g_0(X - ut,u) \propto \exp -[u^2 a + (x - ut)^2/(4a)]. \tag{3.21}$$

If we wish to construct an equation for the marginal distribution $f_t(x)$, note further that $dX_t = U_t dt = U_0 dt$, and hence

$$m(x,t) = E[dX_r | X_r = x]/dt = \frac{E[U_0 X_t]}{E[X_r^2]} x = \frac{t\sigma^2(U_0)}{\sigma^2(X_t)} x$$
$$= tx/(a^2 + t^2),$$

so that

$$\frac{\partial f_t(x)}{\partial t} + \frac{\partial}{\partial x}\left|\left(\frac{tx}{a^2 + t^2}\right) f_t(x)\right| = 0, \tag{3.22}$$

satisfied by $f_t(x)$ in (3.19). Equation (3.22) is not satisfactory, however, as a substitute for (3.20) (or of course (3.17)), as it is no longer a temporally homogeneous Markov process. The same criticism applies to Nelson's more complicated equation (*loc. cit.* p. 139)

$$\frac{\partial f_t(x)}{\partial t} + \frac{\partial}{\partial x}\left| \frac{(t - a)x}{a^2 + a^2} f_t(x)\right| = \frac{1}{2} \frac{\partial^2 f_t(x)}{\partial x^2} \tag{3.23}$$

even though this is also consistent with (3.19) (having been chosen to fit with it).

References

Ballentine, L. E. (1970) The statistical interpretation of quantum mechanics, *Rev. Mod. Phys.* **42**, 358–381.

Bartlett, M. S. (1961) *Probability, Statistics and Time*, Inaugural lecture, University College, London.

——— (1962) Irreversibility and statistical theory, *Essays on Probability and Statistics*, Methuen, London.

Bartlett, M. S. and Moyal, J. E. (1949) The exact transition probabilities of quantum-mechanical oscillators calculated by the phase-space method, *Proc. Camb. Phil. Soc.* **45**, 545.

Bartlett, M. S. (1966) *Introduction to Stochastic Processes*, 2nd ed. Cambridge University Press.

Brillouin, L. (1962) *Science and Information Theory*, 2nd ed. Acad. Press, London and New York.

Broglie, L. de (1964) *The Current Interpretation of Wave Mechanics*, Elsevier, Amsterdam.

Cane, V. R. (1967) Random walks and physical processes, *Bull. Int. Statist. Inst.* **42** (Book 1), 622–39.

Dirac, P. A. M. (1935) *The Principles of Quantum Mechanics*, 2nd ed. Cambridge University Press.

Feynman, R. P. and Hibbs, A. R. (1965) *Quantum Mechanics and Path Integrals*, McGraw, New York.

Gilson, J. G. (1968) Feynman integral and phase space probability, *J. Appl. Prob.* **5**, 375–386.

Kershaw, D. (1964) Theory of hidden variables, *Phys. Rev.* **136B**, 1850–1856.

Landé, A. (1960) *From dualism to unity in quantum physics*, Cambridge.

Mackey, G. W. (1963) *Mathematical Foundations of Quantum Mechanics*, New York.

Marshall, T. W. (1963) Random electrodynamics, *Proc. Roy. Soc.* A **276**, 475–491.

Moyal, J. E. (1949) Quantum mechanics as a statistical theory, *Proc. Camb. Phil. Soc.* **45**, 99–124.

Nelson, E. (1966) Derivation of the Schrödinger equation from Newtonian mechanics, *Phys. Rev.* **150**, 1079–1085.

Nelson, E. (1967) *Dynamical Theories of Brownian Motion*, Princeton Univ. Press, New York.

Varadarajan, V. S. (1962) Probability in physics and a theorem on simultaneous observability, *Comm. Pure Appl. Math.* **15**, 189–217.

Inference and stochastic processes

Reprinted from the *Journal of the Royal Statistical Society, Series A (General)*, Vol. **130**, Pt. 4, (1967) pp. 457–477

SUMMARY

The relation of statistical inference to the wider problem of all inductive inference is reviewed. For scientific inference in general the competing approaches are the hypothetical–deductive and the Bayesian, and the formalism of each is discussed in statistical contexts in terms of the two main concepts of probability—chance and degree of belief.

Inference problems arising with stochastic processes and time-series are considered against this background, and the author's own general attitude to statistical inference reiterated.

Two appendices refer respectively to two specific technical problems (i) separating a discrete and a spectral density component, (ii) specification and inference for "nearest-neighbour" systems.

1. INTRODUCTION

BY the time anyone like myself has been honoured by being elected President of the Royal Statistical Society, he should have become reasonably familiar with the history of his subject; moreover, he will probably have found himself trying to combine with his more routine duties the compilation and presentation of various invited papers and lectures. This creates something of a problem, for just when he would wish to rise to the occasion, he may feel that anything worth while that he might have to say he, or someone else, has said before. In my own case I can but hope that a good deal of the subject-matter of this paper is of sufficient interest to many Fellows for its general lack of novelty, especially in the last few years, to be forgiven. (There is some original material; but this, because of its more specialized character, is relegated to the Appendices.)

This Society has always, and rightly, prided itself on the range of its statistical interests, which extend from the collection and interpretation of statistics for government and commerce to the higher flights of mathematical or philosophical probability. This range of interests does imply, however, a collective responsibility on the part of the Society to maintain a reasonable balance among all the sectional interests there are. In particular, I hope the more mathematical statisticians among us will consider carefully Mr Tippett's general comments in his Presidential Address last year on the importance of balance between statistical theory and practice. The growing need for specialization is in danger of putting us all into little niches which we proceed to blow up in the hope of convincing others that they represent nearly the whole world. Indeed this can be at times, in these days of intense competition between rival activities, almost necessary for adequate survival and development; but I hope we do not really deceive ourselves.

I can imagine that the professional philosophy of a statistician might somewhat differ if he is working in business or in traditional scientific circles, but there I am thinking more of controversies over subjective or objective attitudes to probability, or whether to use decision theory; on the question of theory and practice, I cannot believe that anyone would challenge the scientific view of the essential role of both theory and practical investigation or experiment.

2. Statistical Inference and Scientific Inference

While still on the defensive about my own little niche (which at this stage of my life must be, I suppose, classified as "academic", though, perhaps fortunately for me, this has not always been so), I must be cautious about speaking of problems of inference commonly referred to in textbooks of mathematical statistics as though these are the only problems of statistical interpretation there are. In so far as I have a coherent philosophy of statistics, I hope it is "robust" enough to cope in principle with the whole of statistics, and sufficiently undogmatic not to imply that all those who may think rather differently from me are necessarily stupid. If at times I do seem dogmatic, it is because it is convenient to give my own views as unequivocally as possible.

In previous discussion I have more than once referred to the problems of statistics as falling under the two broad headings of (i) *specification* and (ii) *inference*. What I had in mind was the inclusion under (i) of statistical theories in, say, physics, and the

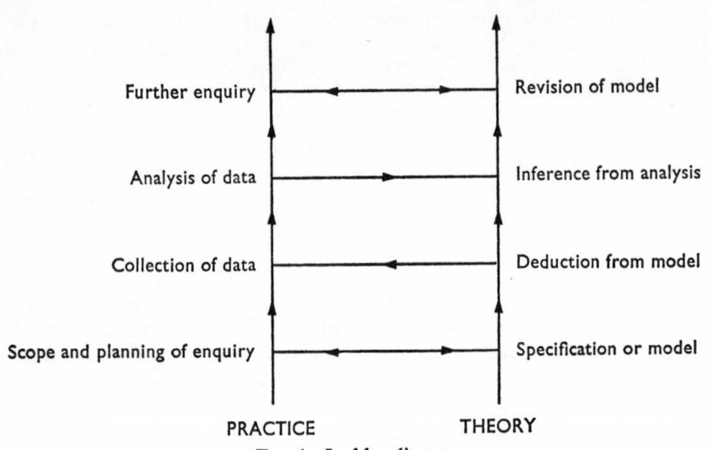

Fig. 1. Ladder diagram.

modern use of more complicated stochastic models in operational research or in biology, as well as the more classical simple assumptions of, say, a normal population or random sampling; and under (ii) the traditional problems of statistical analysis, including estimation, and tests or checks of the specification (I exclude decision theory, as distinct from statistical inference, as not being necessarily *statistical*, in my sense of the word). I still think this demarcation is useful, but I realize it is in danger of underemphasizing the vital point that we must, if we can legitimately call ourselves statisticians, be concerned with real data! Moreover, if I am going to start amplifying, it would seem helpful to expand my (i) and (ii) further, to recognize the increasing need to carry out in (i) rather extensive theoretical investigations, for example, in queueing theory or for epidemic models, before we are in a position to check our specification and in (ii) rather extensive calculations on the data, for example, with the aid of large-scale computers, before we can make our inferences. I should also with this amplification stress somewhere the importance of *planning* the collection of the data, not only in relation to its nature but also to the object of the enquiry. So far my "ladder diagram" (Fig. 1), which hints at the sequential aspect of all scientific enquiry, is not confined to statistical investigations (and I could, if I wanted to, introduce decision theory by substituting *decisions* for *inference*). Statistics is only segregated by the special nature of its subject-matter, namely, observational material classifiable into aggregates or groups. It is this approach which permits, on the practical side, the manageable reduction of suitable data, and, on the theoretical side, the use of statistical probabilities. The possible use of other sorts of probability I will return to, but it is not a question that is the prerogative of the statistician.

My title couples inference and stochastic processes, and I will comment on them together presently, but I have already said that one's approach should be robust enough to cover the entire range of statistics, so that, while special problems of methodology may arise with particular fields, it is quite proper for me to look first at statistical inference in general. It has always been my view that its *theory* should be geared to the concept of statistical probability or chance, and in saying this I do not want now to defend this view at length. It may be queried (see, for example, Bowley's comments on a previous paper of mine to this Society, and my reply (Bartlett, 1940; cf. also Bartlett, 1951), but I regard it as the natural concomitant of the statistical approach, which recognizes systematic variability which is worth segregating, and residual or chance variability which is not. Nothing is claimed about ultimate causes or lack of them, and nothing is claimed about inference methods in general. As I said in my 1940 paper, such a theory may be regarded as a convenient framework on which to fit our facts; and I trust that no one, recalling the development of statistical theory, say, in physics, or in population genetics, is going to query this convenience.

If this is granted, the statistician is in the same position as any other scientist, no worse but no better. Statistical concepts used in his investigations are of use in so far as they lead to better understanding and prediction. Statistical investigations have, however, often to be carried out in rather an empirical manner, for example, in the biological and social sciences, at a stage when comprehensive theories and detailed deductions from them are not yet available. This means that the statistician analysing the results of such investigations uses rather simple or non-specific models, and the problems of analysis and inference take fairly standard and familiar forms. These are the ones to be found in the statistics textbooks. This situation has also encouraged considerable sophistication in the handling of these problems, but I think we must be careful to keep our sense of proportion. Let me, at the risk of monotony, recall what is a basic dilemma. Some philosophers consider that the fundamental problem of inductive inference can be solved by a logical formulation in terms of probability—not statistical probability, but some kind of "degree of belief". Others are equally sure that it cannot. Most scientists sensibly do not allow these continual metaphysical discussions to inhibit their day-to-day work. Unfortunately, the divergence in outlook may noticeably affect the language which statisticians use, much to the confusion of those struggling to understand their discussions. I can thus hardly avoid reminding you of the main points of the two approaches (with their "variations"), and indicate, I am afraid at the cost of considerable repetition of previous remarks, my own standpoint (for a recent historical survey, see also Barnard, 1967).

The method which avoids the acceptance or use of degrees of belief is referred to by Braithwaite (1953) and Popper (1959) as hypothetical–deductive; it is the method which proceeds by the elimination of untenable theories and hypotheses rather than by the selection of the "true", or "most probably true", phrases which some of us regard in many scientific contexts as suspect and unnecessary. The theory of statistical probabilities may be made use of in this general context, and in my opinion it is the parallel between limiting tendencies predictable in the theory (laws of large numbers and of averages, stable distributions and the like) and observed or predicted statistical regularities and frequencies in the real world, that determines broadly the relevance and value of this theory. The status and properties of specific statistical procedures may be examined in detail, but we may claim no more (and no less) than this.

There is one concomitant condition with this approach which Popper (*loc. cit.*, p. 418) rightly emphasizes. The tests and analyses which we perform must be the most severe, and be performed on the most critical type of data, that we can devise. This condition provides the general justification *within this approach* of the various guides which mathematical statisticians are familiar with, including the use of *efficient* estimation, the *most powerful* tests, etc.

Let me now refer to the other approach, associated in recent years with such names as Carnap, de Finetti, Jeffreys and L. J. Savage. Adherents in this country include I. J. Good, D. V. Lindley and C. A. B. Smith.

In this approach a theory of probability is deduced from plausible axioms as a theory of degrees of belief. (The additive law follows if the theory is based on "expected gains", not necessarily otherwise.) The Bayes theorem of inverse probability assumes now a role of fundamental importance, as it permits the derivation of the probability *a posteriori* of a hypothesis subsequent to the consideration of new information S provided the prior probabilities of *all* the possible hypotheses are assessed. A weaker version of this theorem, emphasized, for example, by Good, is merely comparative, and thus avoids the explicit enumeration of all alternatives. It states that the relative odds of two hypotheses H and H' after the consideration of data S are related to the odds prior to the consideration of S by the probability equation

$$\frac{P(H|S)}{P(H'|S)} = \frac{P(H)}{P(H')} \frac{P(S|H)}{P(S|H')}. \tag{1}$$

The relevance of (1) to statistical inference occurs when S denotes statistical data and $P(S|H)$ is identified with a statistical probability $p(S|H)$, say. The second factor on the right-hand side of (1) then becomes the likelihood ratio $\lambda = p(S|H)/p(S|H')$. Equation (1), unlike the absolute form of Bayes's theorem, or the version of (1) involving λ, does not assume the addition rule.

I have repeatedly stressed in previous writings (see, for example, Bartlett, 1965) that it is essential to know what the degrees of belief in this Bayesian theory represent. If they are restricted to situations where they are directly identified with statistical probabilities, no controversy arises. Such situations include some so-called empirical Bayes procedures; and also some applications of decision theory, when the prior probabilities needed may by legitimately interpreted in chance or statistical frequency terms (though this usually means that their precise values are not known).

The interpretation by writers like Good and Savage is that the probabilities are personal assessments. This seems permissible for individual creeds and decisions, but hardly seems acceptable as a basis for scientific objectivity. As Popper (1961, p. 155) puts it: "neither the dryness nor the remoteness of a topic of natural science prevent partiality and self-interest from interfering with the individual scientists' beliefs, and . . . if we had to depend on his detachment, science, even natural science, would be quite impossible." We might somewhat unkindly refer to this theory of inference as "Lamarckian theory", as distinct from a "natural selection theory". Instead of theories surviving because they are the fittest, they do so because someone believes in them hard enough.

This weakness is to some extent avoided by Jeffreys by the substitution of formal or conventional degrees of belief for actual ones. This step has obvious advantages and disadvantages; there is a greater apparent objectivity but also a greater amount of convention. The resulting apparatus may be at times helpful, but its comprehensiveness and availability in general are a matter of opinion. In particular, the prior probabilities of elementary hypotheses, for example, in statistics, may be for convenience allocated simple values (say, in terms of "information theory"), but this is a long way from accepting that the probabilities of all hypotheses can be given numerical values, even conventionally. I have always regarded it as a relevant criticism of this theory that it has been used almost exclusively in statistical situations, where statistical probabilities have been available to give some semblance of reality to the numerical probabilities set down; similarly, my own suspicion of the theory is linked with the consequent degeneration of possible well-authenticated statistical probabilities into formal "degrees of belief". An example given by Popper is of interest, though I do nor regard it as so necessarily fatal to this apparatus as Popper seems to. He cites an imaginary investigation where a Bayesian assigns a prior probability of $\frac{1}{2}$ for getting a head with a toss of a coin, and then proceeds to toss the coin some millions of times, getting exactly 50 per cent of heads. Since the Bayesian's final assessment of the probability remains unaltered, Popper notes that to the Bayesian the information accumulated has been irrelevant, a paradox that in his view raises doubts on the measurement of degrees of belief on a linear scale. I think there is a confusion over prior distributions here. The precise initial assessment of $\frac{1}{2}$ must be adjudged inconsistent with the carrying out of such a laborious and time-consuming investigation, *if* it referred to the constant *chance* of getting heads in such repeated trials; if, on the other hand, it merely referred to the prior distribution of heads and tails at the next throw, then the further trials really were irrelevant.

It seems pertinent to this paradox to recall de Finetti's theorem (cf. Good, 1966) that the consistent assessment of personal probabilities for an observed sequence of, say, heads and tails, provided the permutability condition is satisfied that the probability of the detailed sequence is adjudged a function only of the numbers r of heads and s of tails, say $\phi(r, s)$, must satisfy the equation

$$\phi(r,s) = \int_0^1 p^r(1-p)^s \, dF(p), \tag{2}$$

and is thus consistent with the notion of a probability p with prior distribution $F(p)$. The first case above is the case $F(p) = \delta(p-\frac{1}{2})$, so that

$$\phi(r,s) = \tfrac{1}{2}^{r+s}$$

regardless of what ratio r/s is observed. The second case merely says

$$\phi(1,0) = \tfrac{1}{2} = \int_0^1 p \, dF(p),$$

which is satisfied for any $F(p)$ symmetrical about $p = \frac{1}{2}$.

However, the first case seems to me to emphasize, even after de Finetti's analysis, the distinction between the *believed* value of p, which remained $\frac{1}{2}$ regardless of the observed r/s ratio, and the physical concept which is associated with, and is an idealization of, the r/s ratio (cf. also Hacking, 1965, p. 212). It is possible that for large $r+s$ the latter ratio would be observed to be quite different from $\frac{1}{2}$.

There is thus substance in Popper's general implication that the holding of beliefs and the assessment of information are two very different things; and he has himself made use of formal probability calculus to discuss, not the probability of a hypothesis, but its *degree of corroboration* by data. This appears to me conveniently associated (though not identified) with the difference

$$I(H|S) = \log P(H|S) - \log P(H), \tag{3}$$

which in terms of information theory is the relative information on H conveyed by S (see, for example, Bartlett, 1966a, p. 237; note that, following Barnard, 1951 and Good, 1950, the unaveraged quantity is used, in contrast with, say, Rényi, 1966). While in (3) it depends formally on $P(H)$, it may equivalently be written

$$I(H|S) = \log P(S|H) - \log P(S), \tag{3'}$$

where $P(S)$, while unknown, is perhaps less objectionable. I have purposely not, in (3), used Popper's notation $C(H|S)$ for degree of corroboration, as no absolute measure, only a relative one, is implied by (3). Thus, as Popper notes, if H_0 is another hypothesis entirely independent of H and S, $I(H, H_0|S)$ is the same as $I(H|S)$ in (3), whereas the degree of corroboration $C(H, H_0|S)$ of H and H_0, jointly, by S is intuitively less.

The convenience of (3) or (3′) is that equation (1) now reads

$$I(H|S) - I(H'|S) = \log\{P(S|H)/P(S|H')\}, \tag{1'}$$

where H' is a rival hypothesis to H. In particular, when the right-hand side may be identified with statistical probabilities, we obtain

$$I(H|S) - I(H'|S) = \log \lambda. \tag{4}$$

Notice that (4) is only valid when S is constant for H and H', and even with the identification $P(S|H) = p(S|H)$ we can attach no particular meaning to $\log p(S_1|H) - \log p(S_2|H')$, where S_1 and S_2 are two different samples.

We can, however, write

$$I(H|S_1) - I(H'|S_1) - I(H|S_2) + I(H'|S_2) = \log \lambda_1 - \log \lambda_2, \tag{5}$$

and this may be used to give from (1)

$$\frac{P(H|S_1)}{P(H'|S_1)} = \frac{\lambda_1}{\lambda_2}\,\frac{P(H|S_2)}{P(H'|S_2)}. \tag{6}$$

This does not take us further forward without a new postulate, which appears to have been formulated in different, but related, terms by Barnard (1949) and Hacking (1965). The former writer appeals to a "neutral result" S_2 in relation to the $H : H'$ contrast, the latter to a "principle of irrelevance". To Bayesians this will appear to be cheating and so in a sense it is, for we cannot remove the P's on the right-hand side of (6) (in the absence of *a priori* values) without some identification with the λ's. It is nevertheless true that we may thereby avoid an explicit assumption about $P(H)/P(H')$, and Hacking has made the irrelevance principle the basis for a theory of fiducial probability (cf. also Bartlett, 1966b).

All this is of philosophical interest, but it has no compelling acceptance. Hacking (*loc. cit.*) has made a stimulating attempt to set up a self-contained theory of statistical inference, and he presents it more clearly than Fisher ever did, but his avoidance of

Bayesian arguments on the one hand or "long run" justification on the other seems to me, in spite of my admiration for his attempt, still to leave it somewhat isolated and defenceless. In real life the statistician has the dilemma of having to act both in statistical and non-statistical situations; and he is often accused of being absurd if his statistical inferences are applied to a non-statistical and even unique situation. I have yet, however, to meet any other foolproof procedure in such situations; and if one interprets "long run" justification in a wide enough sense, I still do not know (in spite of Hacking's criticisms) what other acceptable justification there can be. Successful prediction may not be all there is to science, but science could not exist without it. Indeed, one of the virtues of inverse probability and decision theory methods is that when they are applicable to statistical situations with well-defined (frequency) prior distributions and utilities, they lead to the optimum procedure in the statistical sense.

While I would not disagree with Dr Irwin's point (Irwin, 1967) that we employ an act of faith each time we predict into the future, the empirical evidence from the past for learning from experience is otherwise overwhelming (cf. also Hacking, *loc. cit.*, p. 41), and the act of faith can be isolated from the procedure we use. This is perhaps illustrated by the capacity of some computers to learn from experience. Note too that our criterion of their success (say under different program strategies) is of the frequency kind. What criterion do we adopt in assessing the rival merits of, say, a computer as chess player and a human being? Essentially how often one can, or cannot, beat his opponent.

This bears, for example, on my use of (efficient) confidence intervals, in spite of criticism of their relevance, as an *approximation* to the often unattainable ideal procedure. To take a simple example of a normal population with known variance σ^2 and unknown mean μ, the information (in Fisher's sense, but extended to include a known prior normal distribution for μ with variance σ_0^2) is, from a sample of n independent observations, $n/\sigma^2 + 1/\sigma_0^2$: I would, if I ignored the prior distribution, lose that much information. But the Bayesian, if he were wrong about it (in the frequency sense), could lose more; as I have already noted for the binomial sequence, if he were sufficiently pigheaded, the information in the sample might be completely wasted.

3. Stochastic Processes and Time-series

If anyone bothers to check on my attitude to specific "schools" of statistical inference, he may (as Jeffreys once did) accuse me of indefiniteness. My interest in *statistical* phenomena has inclined me to concentrate my attention on the concepts especially useful in analysing and interpreting them, but, like E. S. Pearson and others, I have regarded concepts and techniques as guides rather than rigid recipes. (My complaint with a Bayesian is likely to be more with any apparent intolerance than with his personal creed.) My long-standing interest in stochastic or statistical processes has tended to reinforce this suspicion of over-prescribed disciplines, as is perhaps indicated by my comparatively sparse summary of the principles of statistical inference in my book on *Stochastic Processes* (Section 8.1). In the processes encountered in the real world the extent to which it is conceptually or practically possible to formulate an exact statistical model is very variable, so that precise methods of statistical inference possible in some cases have to be replaced by imprecise and crude methods in others. When confronted with such problems, even setting up the likelihood functions—whether or not to insert into a Bayesian framework—may be a very intractable or dubious procedure (dubious if one knows that the model used is no more than an approximation).

Of course this is not to say that precise methods should not be used when possible; the contributions by Fisher to the design and analysis of statistical experiments are, for example, particularly striking and valuable within their context. In the general field of stochastic processes, the particular domain of stationary time-series stands out as most permanently and completely developed, though the more general paper by Grenander (1950) is an important theoretical landmark.

The analysis of time-series has been of special interest to me from my postgraduate year at Cambridge about 34 years ago, when I first met Udny Yule and found myself struggling with the interpretation of observed serial and cross-correlations. However, as I surveyed the subject of time-series in a paper at the European Meeting of Statisticians last September (see Bartlett, 1967) there is no point in my doing so now in any detail, apart from one outstanding technical point which I have taken the opportunity of discussing further in an Appendix to the present paper. (This concerns the estimation of a spectral density in the neighbourhood of a single discrete spectral component.) Perhaps I should remind you that spectral analysis or the equivalent correlation analysis is based on second-order moment properties, and is thus reasonably exhaustive for normal processes (it will asymptotically specify the likelihood function), but will be less so, like other analysis of variance procedures, in non-normal cases. There is the further problem also, which I can hardly ignore in view of my general theme, of the relevance of sampling properties to our statistical inferences. My view here is of course consistent with my general philosophy that ergodic properties are very relevant, and that we want to have where possible data that we can regard as a reasonably large sample, so that any parametric estimates we make are reasonably accurate. Bayesians—and apparently Barnard, Jenkins and Winsten (1962) in their advocacy of the likelihood function as such—do not always seem to put much emphasis on this requirement; but it is only when we are in the effectively ergodic or large-sample situation that inferences are possible which are not liable to be strongly coloured by the particular approach or views of the individual analysing the data. This result has of course been noted by Bayesians, who demonstrate in their approach the decreasing effect of the prior information. We have seen, however, from de Finetti's theorem for an observed binomial sequence that this is under the conditions that the prior distribution is not extreme but spreads over the entire range, and that the permutability condition is assumed. In that case, even for Bayesians $P\{p-r/(r+s)\to 0\}=1$. I think de Finetti's interpretation of the relation of p to $r/(r+s)$ is illuminating. It contrasts sharply with a pure frequency theory of probability, but, as we have seen, does not imply that the latter *must* be interpreted along subjective lines. Recent work by Kolmogorov and Martin-Löf (see, for example, Marting-Löf, 1966) on the definition and testing of random sequences might in fact be regarded as a partial rehabilitation of the von Mises's approach. Perhaps, as against the subjective approach, only an extreme frequency interpretation, in which chances are *fictitious* idealizations of observed frequencies, will survive; but physical concepts such as space, time, mass, temperature are similarly fictitious and yet invaluable. As I noted in my discussion on Professor Pratt's (1965) paper, such concepts are concrete enough for everyday life. They may all be part of our propensity for model-building, but at least we do seem to have produced *working* models. A stronger light may reveal more clearly how they have been constructed, but this is no reason to abandon them. They may all have their subjective aspects, but it is not such aspects that give them their scientific status.

4. Concluding Remarks on Inference and Stochastic Processes

At one stage in the planning of this Address I entertained the possibility of including some sort of up-to-date survey of inference techniques and problems for stochastic processes, but I soon compromised with this more incomplete discussion. There is now a rapidly growing statistical literature on inference and stochastic processes, not perhaps so vast as the mathematical literature on stochastic processes, but still too large for me to cope with. I merely note, very much at random, Cox's (1966) work on statistical analysis connected with congestion, further work by Curl (1966) on his stochastic models for caves, (cf. Bartlett, 1961 p. 10), and among recent work on the spectral analysis of point processes some in the geophysical domain by Vere-Jones and Davies (1966).

In addition to these very incidental remarks and my first Appendix note, I have included some notes in a second Appendix on inference problems for "nearest-neighbour systems". These are connected with the problems of inference that arise in spatial problems of one or more dimensions. Whittle (1954) has done some pioneering work on such problems, especially on the spectral and autoregressive aspects of stationary processes. Empirical spectral analysis has the attraction of being similar in principle in one or more dimensions, but inference problems for parametric models include some more subtle estimation problems. My own notes refer more to qualitative data—for example, in observations of contagion between arrays of plants or other spatial populations. Spectral or correlation analysis may sometimes be useful here, as I have noted in the case of continuous spatial dimensions (Bartlett, 1964); but for discrete spatial coordinates methods extending probability chain techniques previously developed (see, for example, Bartlett, 1966a, Section 8.2) may be more appropriate.

In view of the somewhat unsystematic character of much of this paper, I recapitulate by way of conclusion my own philosophy of inference. This distinguishes the narrower field of statistical inference from the wider one of inference in general. The former deals with statistical phenomena, and makes use of theoretical (and hypothetical) specifications based on chance. Such methods (including statistical experimentation, with randomization when possible, likelihood ratios, estimation of parameters, tests of significance, confidence intervals) comprise a precise set of *aids* to inference, but not more than that; they have to be considered also in a wider context. In this latter, which we may be more aware of with problems of inference for stochastic processes, I consider that all-embracing formal rules can mislead as much as they simplify. In particular, whenever Bayesian formalism is used, its interpretation should be made unambiguous; in my view it is most acceptable when envisaged in conceptually verifiable (i.e. frequency) terms.

By way of further background emphasis for some of these remarks, perhaps I could end by quoting from an earlier review (Bartlett, 1954). When discussing some of the problems of analysing epidemiological data, I said (p. 130): "It seems likely . . . that statistical inference in the precise sense will remain associated with local or isolated data, such as households or schools, and with the aid of such results available agreement between observation and theory over wider domains would be assessed in a more qualitative manner. This seems quite consistent with scientific method in general. After all, in some fields, such as the theory of evolution, actual measurements are necessarily confined to local or temporary situations or experiments, such as investigating mutation rates, local competition and survival, etc. The statistical analysis of stochastic processes is in any case likely to show a considerable range of approach owing to its close interrelation with the theoretical specification or model an appropriate choice for which becomes of far greater importance in the inference problem for stochastic processes . . . ."

References

BARNARD, G. A. (1949). Statistical inference. *J.R. Statist. Soc.* B, **11**, 115–149.

—— (1951). The theory of information. *J.R. Statist. Soc.* B, **13**, 46–64.

—— (1967). The Bayesian controversy in statistical inference. *J. Inst. Actuar.*, **93** 229.

BARNARD, G. A., JENKINS, G. M. and WINSTEN, C. B. (1962). Likelihood inference and time series. *J. R. Statist. Soc.* A, **125**, 321–372.

BARTLETT, M. S. (1940). The present position of mathematical statistics. *J. R. Statist. Soc.*, **103** 1–29.

—— (1951). Some remarks on the theory of statistics. *Trans. Manchester Statist. Soc.*, Session 1950–51, 1–27.

—— (1954). The statistical analysis of stochastic processes. *Colloque sur l'analyse stochastique Brussels.*

—— (1961). Probability, statistics and time. (Inaugural lecture, University College, London.)

—— (1964). The spectral analysis of two-dimensional point processes. *Biometrika*, **51**, 299–311

—— (1965). R. A. Fisher and the last fifty years of statistical methodology. *J. Amer. Statist. Ass.*, **60**, 395–409.

—— (1966a). *An Introduction to Stochastic Processes* (2nd ed.). Cambridge: University Press.

—— (1966b). Review of *Logic of Statistical Inference* by I. Hacking. *Biometrika*, **53**, 631–633.

—— (1967). Some remarks on the theory of time-series. *Biometrika*, **54**, 25–38.

BRAITHWAITE, R. B. (1953). *Scientific Explanation.* Cambridge: University Press.

BROOK, D. (1964). On the distinction between the conditional probability and the joint probability approaches in the specification of nearest-neighbour systems. *Biometrika*, **51**, 481–483.

COX, D. R. (1958). The regression analysis of binary sequences. *J. R. Statist. Soc.* B, **20** 215–242.

—— (1966). Some problems of statistical analysis connected with congestion. In *Proceedings of the Symposium on Congestion Theory* (ed. W. L. Smith and W. Wilkinson). Chapel Hill University of North Carolina Press.

CURL, R. L. (1966). Caves as a measure of Karst. *J. Geol.*, **74**, 798–830.

GHENT, A. W. (1963). Studies of regeneration in forest stands devastated by the spruce budworm III: Problems of sampling precision and seedling distribution. *Forest Sci.*, **9**, 295–310.

GHENT, A. W and GRINSTEAD, B. (1965). A new method of assessing contagion, applied to distribution of red-ear sunfish. *Trans. Amer. Fisheries Soc.*, **94**, 135–142.

GOOD, I. J. (1950). *Probability and the Weighing of Evidence.* London: Griffin.

—— (1966). How to estimate probabilities. *J. Inst. Math. Applics.*, **2**, 364–383.

GRENANDER, U. (1950). Stochastic processes and statistical inference. *Ark. Mat.*, **1**, 195–277.

HACKING, I. (1965). *Logic of Statistical Inference.* Cambridge: University Press.

IRWIN, J. O. (1967). William Allen Whitworth and a hundred years of probability. *J. R. Statis Soc.* A, **130**, 147–175.

KENDALL, M. G. (1966). Statistical inference in the light of the theory of the electronic computer *Rev. Int. Statist. Inst.*, **34**, 1–12.

MARTIN-LÖF, P. (1966). Algorithms and randomness. (Paper presented to the European Meeting of Statisticians, London.)

PLACKETT, R. L. (1966). Current trends in statistical inference. *J. R. Statist. Soc.* A, **129** 249–267.

POPPER, K. R. (1959). *The Logic of Scientific Discovery.* London: Hutchinson.

—— (1961). *The Poverty of Historicism.* London: Routledge & Kegan Paul. (Paperback edition

PRATT, J. W. (1965). Bayesian interpretation of standard inference statements. *J. R. Statist. Soc.* B, **27**, 169–203.

RÉNYI, A. (1966). Statistics based on information theory. (Paper presented to the European Meeting of Statisticians, London.)

SEAL, H. L. (1966). Testing for contagion in animal populations. *Trans. Amer. Fisheries Soc.*, **95** 436–437.

THOMPSON, H. R. (1955). Spatial point processes, with applications to ecology. *Biometrika*, **42** 102–115.

VERE-JONES, D. and DAVIES, R. B. (1966). A statistical survey of earthquakes in the main seismic region of New Zealand. Part 2: Time series analyses. *New Zealand J. Geol. and Geophys.*, **9**, 251–284.
WHITTLE, P. (1954). On stationary processes in the plane. *Biometrika*, **41**, 434–449.
—— (1963). Stochastic processes in several dimensions. *Bull. Int. Statist. Inst.*, **40**, 970–994.

APPENDIX I

Use of Phase in Separating a Discrete and a Spectral Density Component

The amplification of the proposal included in my survey of time-series at the European meeting of Statisticians (cf. Bartlett, 1967) is given below. The asymptotic relation was given

$$J'_Y \sim J_Y + \alpha\sqrt{(\tfrac{1}{2}n)}\, e^{-i\Phi}\frac{\exp(2\pi i\epsilon)-1}{2\pi i(\epsilon+s)}, \tag{1}$$

where

$$Y'_r = Y_r + \alpha\cos(rv+\Phi),$$

$$J_Y = \sqrt{(2/n)}\sum_{r=1}^{n} Y_r \exp(ir\omega_p),$$

$$\omega_p = 2\pi p/n, \quad v = \omega_{p-s} + 2\pi\epsilon/n \quad (-\tfrac{1}{2}<\epsilon<\tfrac{1}{2}),$$

Φ is a random phase angle.

In an example of the spectral analysis of traffic data (time instants of successive vehicles passing a fixed point on a road) referred to in (Bartlett, 1966a, p. 330) the estimate 1·849 was obtained from the mean of the outlying 18 unsmoothed periodogram intensities JJ^*, the central 16 values not being used because of the presence of one or more anomalous values to be tested. (These data relate to *point processes*, but this aspect may be ignored for the present discussion.) The individual A_p and B_p values, where $A_p + iB_p = J(\omega_p)$, are shown in Fig. 2.

A rapid way of obtaining further information on the spectral density is to measure deviations orthogonal to the fitted line corresponding to the relation (1) above. In the case of one intensity much larger than the others, it will usually be sufficient to fit the line by this observation alone. This prodecure results in the estimate 1·849 (36 d.f.) being modified to

$$\frac{33{\cdot}29+16{\cdot}23}{18+\tfrac{1}{2}(15)} = 1{\cdot}942 \quad (51 \text{ d.f.}).$$

In principle, the information in the deviations along the fitted line is also recoverable, though more cumbersome to extract. (The entire procedure for the recovery of this further information on the spectral density depends of course on the supposition that only *one* discrete component is present in the interval considered, as has been assumed by relation (1).) From the form of equation (1), we shall minimize.

$$S = \textstyle\sum_s\{(A_s-\beta a_s)^2 + (B_s-\gamma a_s)^2\},$$

where $a_s = 1/(\epsilon+s)$, β and γ are also functions of ϵ, and it is assumed that the axes have been rotated so that the estimate $\hat{\gamma} = 0$. We obtain

$$\begin{aligned}
\partial S/\partial\beta &= \textstyle\sum_s(A_s-\beta a_s)\,a_s = 0,\\
\partial S/\partial\gamma &= \textstyle\sum_s(B_s-\gamma a_s)\,a_s = 0,\\
\partial S/\partial\epsilon &= \textstyle\sum_s(A_s-\beta a_s)\,(\beta\partial a_s/\partial\epsilon + a_s\,\partial\beta/\partial\epsilon)\\
&\quad + \textstyle\sum_s(B_s-\gamma a_s)\,(\gamma\partial a_s/\partial\epsilon + a_s\,\partial\gamma/\partial\epsilon)\\
&= \textstyle\sum_s(A_s-\beta a_s)\,\beta\partial a_s/\partial\epsilon = 0,
\end{aligned}$$

the last equation simplifying by use of the first two. From the equation $a_s = 1/(\epsilon+s)$, the last equation may be written

$$\Sigma_s(A_s - \beta a_s)\, a_s^2 = 0.$$

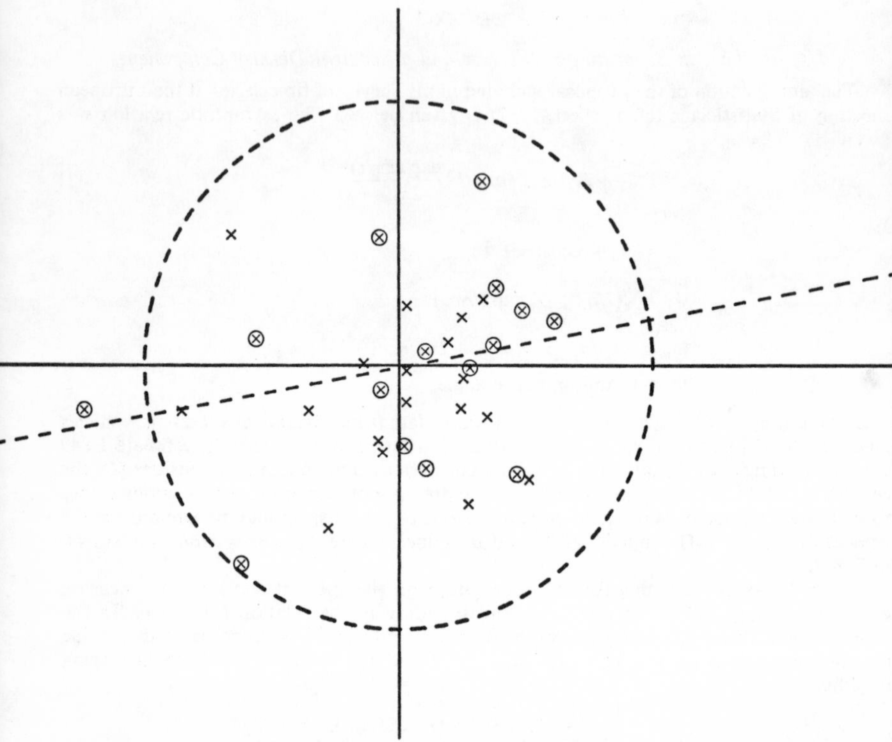

FIG. 2. Spectral analysis example. Thirty-four individual A_p and B_p plotted as abscissae and ordinates. Ringed points refer to the central 16 values, and the dotted circular boundary denotes the $P = 0{\cdot}01$ significance level for the individual points.

This equation, in conjunction with the equation for β, was used to obtain the estimate for ϵ by interpolation

$$\hat{\epsilon} = 0{\cdot}368,$$

giving an estimate of the corresponding period of $1700/20{\cdot}368 = 83{\cdot}46$ seconds. This period corresponded to a known period given as 83 seconds due to traffic lights "upstream". As the approximate standard error is 0·29 seconds, obtained by a consideration of the effect of the random fluctuations in $J'_s = A'_s + iB'_s$, this estimate appears quite satisfactory.

Finally, the value of S was used to obtain the most efficient estimate of the spectral density. The final value obtained was 1·830 (66 d.f.), compared with the original value 1·849 (36 d.f.), and the intermediate value 1·942 (51 d.f.) which used only half the further information.

APPENDIX II

Specification and Inference for Nearest-neighbour systems

1. *Simultaneous and Conditional Probability Systems*

Whittle (1963) and Brook (1964) have noted that some care is necessary in specifying a nearest-neighbour system. Either this may be done in terms of conditional probabilities, a practice I have tended to adopt (cf. Bartlett, 1966a, Section 2.22) or in terms of simultaneous probabilities. With the former definition we assume

$$p(X_j \mid X_{j+1}, X_{j+2}, \ldots, X_{j-1}, X_{j-2}, \ldots) = p(X_j \mid X_{j+1}, X_{j-1}), \tag{1}$$

and with the latter

$$p(\ldots, X_{j-2}, X_{j-1}, X_j, X_{j+1}, X_{j+2}, \ldots) = \Pi_j\, Q_i(X_{j-1}, X_j, X_{j+1}). \tag{2}$$

This last equation is the "open" form given by Whittle. Brook, however, gives the "closed" form

$$p(X_1, X_2 \ \ldots, X_n \mid X_0, X_{n+1}) = \prod_{j=1}^{n} Q_j(X_{j-1}, X_j, X_{j+1}) \tag{3}$$

and notes an example, viz.

$$Q_j(X_{j-1}, X_j, X_{j+1}) = (X_{j-1} + X_{j+1}) \exp\{-X_j(X_{j-1} + X_{j+1})\}\, dX_j, \tag{4}$$

which is not reducible to (1). It is not clear whether (3) is supposed to be valid for fixed n (and the origin) or arbitrary n, and difficulties may arise in the second case. We could then, for example, deduce immediately ($n = 1$)

$$p(X_1 \mid X_0, X_2) = Q_1(X_0, X_1, X_2), \tag{5}$$

and while this does not imply (1) it implies severe limitations on admissible systems unless the Q_j became very artificial. In particular, (1) is not included very easily under (3). For (1) implies a Markov chain, so that

$$\begin{aligned} p(X_1, X_2, \ldots, X_n \mid X_0, X_{n+1}) &= \frac{p(X_1, X_2, \ldots, X_{n+1} \mid X_0)}{p(X_{n+1} \mid X_0)} \\ &= \left\{\prod_{j=1}^{n} p(X_j \mid X_{j-1})\right\} \frac{p(X_{n+1} \mid X_n)}{p(X_{n+1} \mid X_0)} \\ &= \prod_{j=1}^{n} \left\{\frac{p(X_j \mid X_{j-1})\, p(X_j \mid X_0)}{p(X_{j+1} \mid X_0)}\right\} \end{aligned}$$

This is indeed of the form (3) if X_0 and X_{n+1} are considered fixed, but the identification (5) does not extend to the other Q_j. On the other hand, with (2), we have for a Markov chain

$$p(\ldots, X_{j-2}, X_{j-1}, X_j, X_{j+1}, X_{j+2}, \ldots) = \ldots p(X_{j-1} \mid X_{j-2})\, p(X_j \mid X_{j-1})\, p(X_{j+1} \mid X_j) \ldots$$

which is of the required form.

More non-degenerate models of (3) (and their generalizations in more dimensions) would be useful, so that their value, for example, for inference purposes, could be assessed. It is possible to postulate equilibrium chains arising from "potential functions" as in statistical mechanics, but even then interaction between neighbouring *pairs* would seem more natural. In any case, what I have in mind are models arising from a detailed stochastic mechanism; without this there is always the risk of inconsistent or degenerate models. So far I have not succeeded in inventing one that satisfies (3).

Superficially, if equation (3) were available estimation and goodness-of-fit techniques would be immediately available as in Markov chain and process problems; for example, if the Q functions existed and were known for a probability chain with a finite number of discrete states. (The form of (3) would imply central limit properties under fairly wide conditions.) However, if the Q functions are not known their non-identification with conditional probabilities implies, even in, say, the discrete probability chain case, no means of estimating probabilities over *subsets* of the observations, and this seems to rule out simple goodness-of-fit tests. This difficulty is illustrated further below.

Estimation methods seem straightforward. Suppose, for example, in place of (4) we included a sequence with unknown scaling factor θ, so that

$$Q_j(X_{j-1}, X_j, X_{j+1}) = (X_{j-1}+X_{j+1}) \exp\{-X_{j-1}+X_{j+1})/\theta\}\, dX_j/\theta,$$

Then for the log likelihood function

$$L = \textstyle\sum_j \log(X_{j-1}+X_{j+1}) - X_j(X_{j-1}+X_{j+1})/\theta - \log\theta,$$

$$\frac{\partial L}{\partial\theta} = \textstyle\sum_j \left\{\frac{X_j(X_{j-1}+X_{j+1})}{\theta^2} - \frac{1}{\theta}\right\}$$

$$= \frac{n}{\theta^2}\left\{\frac{\sum_j X_j(X_{j-1}+X_{j+1})}{n} - \theta\right\}. \qquad (6)$$

This implies (if the information in X_0 and X_{n+1} is considered unavailable), the Fisher information function n/θ^2, and the further differentiation of L leads to asymptotic confidence intervals for θ just as in the one-sided case (see, for example, Bartlett, 1966a, p. 254). It might be noted that the estimation of θ depends on the sufficient statistic $\sum_j X_j(X_{j-1}+X_{j+1})/n$, and this for large n will not differ appreciably from $2\sum_j X_j X_{j-1}/n$, a statistic relevant to the one-sided case. Thus even in this example, as far as estimation is concerned, the two-sided character of the probabilities does not seem to lead to anything very different from the one-sided case, in spite of its formal incompatibility with it.

2. *Goodness-of-fit Tests*

The extension of goodness-of-fit techniques to nearest-neighbour systems is discussed in relation to one or two examples. The first one-dimensional one is unsatisfactory because of the absence of any simple mechanism for generating Q-chains (as I shall call them), so that in order to obtain a nearest-neighbour type system which did not degenerate to a Markov chain I first took a random sequence of 102 0's and 1's and then allowed the 1's (as if infected plants in a row) to produce 1's on either side with a probability of $\frac{1}{5}$ if the individual item in question were previously a 0. The resulting overall probability of a 1 for any item then becomes $\frac{1}{2}+\frac{1}{2}\cdot\frac{1}{2}\cdot\frac{1}{5}+\frac{1}{2}\cdot\frac{1}{2}\cdot\frac{1}{5}-\frac{1}{2}\,\frac{1}{2}\,\frac{1}{2}\,\frac{1}{25} = 0{\cdot}595$, the last correcting term arising from the chance of a 0 being simultaneously "infected" from both sides. The sequence so obtained is given below:

1, 10110, 01100, 00001, 11110, 00111, 10101, 11110, 11001, 10111, 01011,

10111, 00011, 01110, 10000, 00000, 11011, 11110, 11001, 11011, 00100, 0.

The frequencies $n(x_{i-1}, x_i, x_{i+1})$ of different triads (x_{i-1}, x_i, x_{i+1}) are shown in Table 1, with the expected frequencies, on the *null* (and incorrect) hypothesis of independent probabilities $p(0) = 0{\cdot}405$, $p(1) = 0{\cdot}595$, given also in brackets.

TABLE 1

1	000	14 (6·64)	5	010	4 (9·76)
2	001	8 (9·76)	6	011	17 (14·34)
3	100	9 (9·76)	7	110	18 (14·34)
4	101	13 (14·34)	8	111	17 (21·06)

The formal chi-square is 14·27, which we might at first perhaps be tempted to assign 7 d.f. However, suppose we had intended to examine the null hypothesis in relation to the class of Markov chains of dependence two; our classification of the above data would then have been as in Table 2 (cf. Bartlett, 1966a, p. 263).

TABLE 2

	00	01	10	11	Total
00	14 (6·64)	.	9 (9·76)	.	23
01	8 (9·76)	.	13 (14·34)	.	21
10	.	4 (9·76)	.	18 (14·34)	22
11	.	17 (14·34)	.	17 (21·06)	34
Total	22 (16·4)	21 (24·1)	22 (24·1)	35 (35·4)	100

We now obtain $\chi^2 = 14{\cdot}27 - 2{\cdot}50 = 11{\cdot}77$, with 4 d.f., so that the null hypothesis would be correctly rejected ($P \sim 0{\cdot}02$). This calculation reminds us, however, of the equality (or near equality) of the corresponding row and column totals, a set of constraints we might have overlooked for the Q-chain model, as the corresponding likelihood theory merely tells us that $Q(x_{i-1}, x_i, x_{i+1}) \propto n(x_{i-1}, x_i, x_{i+1})$. It is easy to check from its method of construction that neither the Markov chain nor Q-chain model is theoretically valid for this example, but it has served to illustrate difficulties with the Q-chain model, in spite of the latter's apparent simplicity. It should be emphasized that with *conditional* probability systems, the evaluation of the *absolute* expected frequencies is unnecessary (see equation (6), Chapter 8.2, Bartlett, 1966a): and, if they are introduced, this is purely for convenience or interest.

In the two-dimensional case I shall therefore concentrate on the conditional probability procedure. Consider a one-sided Markov type dependence extending to the top and left of each item, so that with the top row and column assumed given, and a specified conditional probability of 0 or 1 for any item, we may generate a whole rectangle of observed values. The probability-scheme was as shown in Table 3. This scheme is for simplicity somewhat degenerate, being specified by

TABLE 3

	0 0 .	1 0 .	0 1 .	1 1 .
0	0·8	0·5	0·5	0·2
1	0·2	0·5	0·5	0·8

$$p(x_{ij}=1 \mid x_{i-1,j}, x_{i,j-1}) = 0{\cdot}2 + 0{\cdot}3(x_{i-1,j} + x_{i,j-1}).$$

Averaging over all x, we obtain for the overall probability p per item

$$p = 0{\cdot}2 + 0{\cdot}3(p+p), \tag{7}$$

whence $p = \frac{1}{2}$. Again for convenience, the first row and column consisted of *random* 0's and 1's. An artificial array of $21 + 100$ items so obtained is given in Table 4.

The observed (and expected) frequencies corresponding to the above probability scheme are given in Table 5.

In order to obtain the theoretical expectations on the true scheme (as distinct from the null hypothesis of a random scheme), we note that

$$p(x_{ij}=1 \mid x_{i-1,j}, x_{i,j-1}) = E\{x_{ij} \mid x_{i-1,j}, x_{i,j-1}\}.$$

Hence equation (7) above gives the regression equation of x_{ij} on $x_{i-1,j}$ and $x_{i,j-1}$. Let

$$m_{rs} = E\{x_{ij}\, x_{i-r,j-s}\}.$$

TABLE 4

1	0	1	0	1	0	0	0	1	1	0
0	0	0	0	0	1	0	0	1	1	0
0	0	0	1	1	0	0	1	1	1	0
0	0	0	1	0	1	0	1	1	0	0
0	0	0	1	1	1	0	1	1	1	1
1	1	1	0	0	1	1	0	1	0	0
0	1	1	0	0	0	1	1	0	0	1
0	1	1	0	0	1	1	1	0	0	0
1	1	1	0	0	1	1	1	0	0	0
0	1	1	1	1	1	1	1	1	0	0
1	1	1	1	1	0	0	1	1	0	0

TABLE 5

	0 0 .	1 0 .	0 1 .	1 1 .	Total
0	20 (24·4)	7 (9·75)	12 (9·75)	8 (6·1)	47
1	6 (6·1)	13 (9·75)	11 (9·75)	23 (24·4)	53
Total	26 (30·5)	20 (19·5)	23 (19·5)	31 (30·5)	100

Multiply equation (7) by $x_{i,j-1}$ or $x_{i-1,j}$ and average. This provides the relation

$$m_{10} = m_{01} = 0{\cdot}1 + 0{\cdot}3(m_{-1,1} + 0{\cdot}5)$$

or

$$m_{10} = m_{01} = 0{\cdot}25 + 0{\cdot}3m_{-1,1} \tag{8}$$

where the identity $x_{i,j-1}^2 = x_{i,j-1}$ has been used. To obtain a further relation, note that

$$p(x_{ij} \mid x_{i,j-1}, x_{i-1,j-1}, x_{i-2,j}) = 0{\cdot}2 + 0{\cdot}3\{x_{i,j-1} + 0{\cdot}2 + 0{\cdot}3(x_{i-1,j-1} + x_{i-2,j})\}.$$

Multiply by $x_{i-1,j-1}$ and average, and we obtain

$$m_{11} + 0{\cdot}13 + 0{\cdot}3m_{10} + 0{\cdot}09(m_{-1,1} + 0{\cdot}5),$$

or (if we assume $m_{-r,s} = m_{rs}$)

$$0{\cdot}91m_{11} = 0{\cdot}175 + 0{\cdot}3m_{10}. \tag{9}$$

From (8) and (9), we obtain $m_{11} = 0{\cdot}305$.

Hence, using also the symmetry relations between 0 and 1, we obtain the expected values given in Table 3. The likelihood argument gives a χ^2 formula exactly analogous to the one-dimensional Markov chain case, this because of the *conditional probability* model used. We thus obtain a test of the true model

$$\chi^2 = 4{\cdot}00 - 1{\cdot}31 = 2{\cdot}69,$$

with 4 d.f., this value being satisfactory.

In this example the absolute expected frequencies were calculated for interest, though on a rather *ad hoc* basis (notice that while equation (7) has the character of a regression equation we cannot use Whittle's solution (1954, p. 447) as the conditional variance of x_{ij} is not independent of $x_{i-1,j}$ and $x_{i,j-1}$). In more complex problems this calculation would be best avoided by the direct and simpler use of the conditional distributions.

If a more general theory of the probabilities were needed, applicable to any finite number of states and conditional distributions, it might perhaps be developed as an example of the theory of *random operators*. In the present context we have a sequence of states determined not only necessarily by the preceding state, but by the corresponding one on the row above. The effect of the latter is to alter the transition probability matrix operating in the Markov chain sense, so that we have what might be termed a *doubly stochastic Markov chain*. The order of the sequence of different matrices is determined by the realization of states which in turn were the outcome of the previous row's sequence of matrices.

TABLE 6

Number of seedlings in five feet square quadrats

(Taken from Ghent, 1963, Fig. 4.)

0	1	2	3	4	3	4	2	2	1
0	2	0	2	4	2	3	3	4	2
1	1	1	1	4	1	5	2	2	3
4	1	2	5	2	0	3	2	1	1
3	1	4	3	1	0	0	2	7	0
4	2	0	0	2	0	3	2	3	2
2	2	2	0	3	4	7	4	3	3
2	3	1	2	3	8	5	5	1	2
1	1	2	1	4	4	5	3	2	3
3	1	6	1	3	5	4	7	4	3

The above work is relevant to a recent discussion (Ghent and Grinstead, 1965; Seal, 1966) on testing for contagion. It indicates in particular valid extensions of the chi-square test for some nearest-neighbour systems in two (or more) dimensions along lines previously developed by the author for Markov chains in one dimension. (In the two-dimensional case a rectangular grid of points has been assumed, but a more irregular boundary could be acceptable provided the conditional distributions specifying the model are realizable by a sufficiently extensive given set of boundary points.)

In an earlier paper (1963), Ghent gives an example (his Fig. 4, p. 302) of balsam-fir seedling counts. Various analyses would be possible of these data, including a spectral analysis on the detailed coordinates (see Bartlett, 1964) and a straightforward nested analysis of variance on the counts in five feet square quadrats, given in Table 6.

The latter dispersion analysis (cf. Thompson, 1955) within and among sets of four gives the χ^2 figures of Table 7. This is quite sufficient, in spite of the non-significant total dispersion noted in effect by Ghent, to reject the null hypothesis of randomness from the χ^2 item among sets of four ($P \sim 0{\cdot}001$). Any more complicated analysis would hardly be justified without further attention to the ecological situation, but as Ghent (see his Fig. 7, p. 306) classifies the counts of Table 6 above in three categories—low density (0 or 1), medium (2 or 3) and high (4 or more)—it is perhaps of interest to illustrate the "nearest-neighbour" χ^2 technique on these data, taking as before "one-sided dependence" and the top row and left column as given. The following frequency table (Table 8) was obtained, where low density is denoted by L, medium by M and high by H.

To check the null (random) hypothesis, the conditional distributions are taken constant and estimated by the row totals. This is equivalent to calculating the usual contingency table χ^2, which is 23·36 with 16 d.f. This value is not significant ($P \sim 0{\cdot}10$), but if we group on the basis of the total score, taking $L \equiv -1$, $M \equiv 0$, $H \equiv +1$, we get the reduced Table 9 for which the contingency table χ^2 is 19·68 with 8 d.f. ($P \sim 0{\cdot}01$).

TABLE 7

χ^2 analysis of seedling data

	D.f.	χ^2
Among sets of four	24	52·4
Within sets of four	75	65·6
Total	99	118·0

TABLE 8

	L L .	L M .	M L .	M M .	H M .	M H .	H H .	H L .	L H .	Total
L	2	5	6	5	1	2	0	0	3	24
M	6	3	4	9	3	4	3	2	2	36
H	0	3	4	1	3	4	5	1	0	21
Total	8	11	14	15	7	10	8	3	5	81

To fit a more adequate model, let us suppose that the L (or -1) and H (or $+1$) probabilities regress linearly on the conditioning total score of Table 9 (cf. the previous artificial example). To keep the model as simple as possible, we assume provisionally that the "regression" for the central M (or 0) value is zero. We then have

$$p(x_{ij} = r \mid x_{i-1\,j}\, x_{i,j-1}) = \alpha_r + \beta rs, \tag{10}$$

where $s = x_{i-1,j} + x_{i,j-1}$. The model is obviously rather empirical; and while it would not be difficult to set out the maximum likelihood equations for α_r and β, they are less tractable than calculating the unweighted "regression" estimate for β which is 0·107. The expected frequency values with this value for β are given in brackets in Table 9, and the amended χ^2 value is now 11·94 with 7 d.f., giving an improved fit ($P \sim 0{\cdot}10$).

TABLE 9

r \ s	−2	−1	0	1	2	Total
−1	2 (4·00)	11 (9·82)	8 (6·58)	3 (3·04)	0 (0·58)	24
0	6 (3·55)	7 (11·10)	13 (10·21)	7 (7·55)	3 (3·55)	36
1	0 (0·45)	7 (4·07)	2 (6·21)	7 (6·41)	5 (3·87)	21
Total	8	25	23	17	8	81

This fit could perhaps be improved further either by the use of a more efficient estimate for β, or by the choice of a more sophisticated model, for example, of the logistic type used by Cox (1958) in the case of binary sequences.

I am indebted to Mr R. Galbraith for checking through my calculations in Appendix II.

INFERENCE AND STOCHASTIC PROCESSES

Proceedings of the Meeting

Mr L. H. C. Tippett: As a fellow ex-Mancunian, I am specially glad to have the honour of proposing a vote of thanks to our President. In many ways Manchester continues to be something of a pioneer and its university was very early, if not first, in the post-war expansion of statistics in establishing a chair of mathematical statistics, to which it appointed Maurice Bartlett. Later, in 1959, the Manchester Statistical Society anticipated the present occasion by electing him its President. Subsequently, of course, Professor Bartlett moved to London and shortly, as we know, is to move to Oxford.

Professor Bartlett accepted the office of President of this Society at a very busy time for him. He had only just completed a term as President of the British Region of the Biometric Society and was in the middle of a term as President of the International Association for Statistics in Physical Sciences, which is a section of the International Statistical Institute. He is active with the Royal Society—I believe he is a member of its council—and in addition has all the administrative duties that seem almost to submerge professors of statistics in these days. Nevertheless, he has given full attention to the affairs of our Society and, as the Annual Report states, his many contacts have been much appreciated. He has long been an active member of our Society and his influence will continue to be felt and to be valued in the future. We should thank Professor Bartlett for his work for the Society.

I also propose that we thank him for his Address. You will be reassured to know that the Editor of Series A likes it and will be glad to see it in the *Journal*. It would be an interesting situation if the Editor wanted to reject the Presidential Address, wouldn't it? Professor Bartlett need not apologize for the "general lack of novelty", as he terms it, of much of the subject-matter. Many Fellows will be glad to read an up-to-date survey of problems of inference and a statement of Professor Bartlett's own position. His contributions to the science of statistics are many and distinguished. His theoretical work, notably in stochastic processes and multivariate analysis, is important. And he is one of those rarer Fellows who combine eminence in theoretical work with considerable interest and competence in practical application. His published papers cover agriculture, epidemiology, econometrics, physics and other subjects. It is good to have the working philosophy of such a statistician.

It is not clear whether the convention that the Presidential Address shall not be discussed is to protect the President from the kind of rough treatment that this Society sometimes hands out to its paper readers or to prevent embarrassment to proposers of votes of thanks who have been chosen for reasons other than their expertise in the field under discussion in the paper. I certainly am glad to shelter behind the convention, because I always approach a paper such as the Address before us this evening in a spirit of humble, though interested, enquiry.

Precedent, however, does permit me—indeed, may compel me—to present general reflections on the Address, and in doing so I realize that some of them merely restate in other words what the President has said; and I can only hope that the other things that I shall say are not nonsense.

To his two approaches to inference may be added a third, that of people who are content to work by "cook-book" rules without caring much about their mathematical or philosophical justification. These people do not seek optimum decisions or anything of that sort; they merely ask that the rules will keep them out of trouble. They are very numerous and they must be reckoned with.

Statisticians who adopt the President's first approach, of whom I must confess I am one, are only one stage more rational. They accept without question the general practices of scientific inference that they have picked up in the course of their education and experience (they scarcely know how) and merely seek to fit statistical inferences, with probability and all the rest of it, into that scheme. Statistics enables them to decide if any given hypotheses are compatible with the data, but it does not help them to choose between compatible ones.

Statisticians who adopt the President's second approach are really trying, it seems to me, to develop a complete theory of knowledge—a formidable task in which philosophers do not so far seem to have been entirely successful. Subjectivity has not been eliminated from this field, and it is likely that attitudes and conclusions are influenced by temperament and the prejudices collected from life, as well as by reason. Attempts to develop these theories are well worth making and those who make the attempts must believe in what they do. An "apparent intolerance", which the President seems to discover in some Bayesians, is a natural thing. It would be a pity if workers in this field were either discouraged or distressed by the scepticism some of us feel for the validity of their results.

In his contributions to discussions of inference, the President shows to an unusual degree a sympathetic understanding of other people's views, and an ability to discuss them and his own in a sweetly reasonable way. This attitude may open him to the charge of "indefiniteness", which he mentions in the opening sentence of his third section, but it makes him an excellent medium of communication between the various schools of thought and between people of those schools and of no school, and thus he contributes signally to the advancement of knowledge.

I have the honour to propose a vote of thanks to our President for his work for the Society and for his Address.

Dr J. O. Irwin: On a good many occasions in my past lifetime I have been quite used to playing second fiddle, but tonight I am doing so in a double capacity. I do not regard it as at all extraordinary that I should be seconding a vote of thanks proposed by my old friend, Tippett; but I certainly did not think that it would fall to me to understudy the Chairman of Imperial Chemical Industries. I know how much Sir Paul Chambers must regret his inability to be here, especially as our President was for a few years a member of the I.C.I. scientific staff. However, I think I can say that I have known our President longer than Sir Paul and I suspect—I hope the President will not consider this a disadvantage—that I know him better.

Since it is less than 48 hours ago that I had notice of this assignment, I have not been able to refresh my memory as well as I would have wished on his early work and the many things which he has done since, so the President may be assured that any nice things I happen to say about him are pretty spontaneous.

I think we first met in 1932 or 1933; another old friend, Jack Wishart, had gone to Cambridge as reader in statistics, and our President was the first mathematical star that Wishart attracted into the then rather unusual orbit of statistics. It is my belief that even in those early days—for I was looking yesterday at two joint papers by Wishart and Bartlett on the distribution of second-order moment statistics in a normal system—our President gave at least as much as he received. Since then the statistical world has become very much his debtor for his contributions, both to theory and practice.

Before the second war, I can recall early papers on distribution theory, on statistical inference, especially in relation to sufficiency and fiducial probability, on factor analysis, on the meaning of probability, on experimental design and on the χ^2 test for the significance of the difference between a number of variance estimates. In 1941 we had his paper on the statistical significance of canonical correlations. From 1945 onwards, stochastic processes seem to be the *leit-motif* with applications in biology to population growth, population genetics and epidemiology, and in physics to communication theory and other subjects. The President's book on stochastic processes, published in 1955, I would guess (or I might say *στοχάζομαι*) marks as much of an epoch in one direction of progress as exactly 30 years earlier R. A. Fisher's *Statistical Methods for Research Workers* did in another.

In addition to all this, we must remember all the work he has done as consultant, teacher and head of large departments.

Criticism of the Presidential Address is by custom inadmissible. I have given my own views on some of the topics discussed in Section 2 a number of times before, and I have said on more than one occasion that the probabilities I think I am using in statistical investigation are statistical probabilities in the President's sense. It seems to me that this is not the occasion to go beyond this point in discussing statistical and scientific inference. Nor would it be appropriate for me to say anything about the present position of stochastic process theory and techniques. I have studied what Yule and McKendrick did before the name was invented—and this, it is now generally recognized, was not a little—but I am all too ignorant of the later developments.

I have always been fascinated, however, by the word "stochastic". Tschuprow, in his book on correlation theory, written in 1925, in German, speaks of chance variables but of the "stochastic dependence" between them. In a footnote he says the word came from the Greek word *στοχάζομαι*, while he himself uses "*stochastisch*" as synonymous with "*wahrscheinlichkeitstheoretisch*", and he gives a reference to J. Bernoulli's *Ars Conjectandi*, p. 213 (Basle, 1713), and the work of L. von Bortkiewiscz on runs, *Die Iterationen* (Berlin, 1917). I do not know who first used the term "stochastic variable" but I could not resist looking up the word *στοχάζομαι* in Liddell and Scott. There I found it went back before Plato and was used by a number of Greek writers, including Aristotle and Galen. It seems to have had two main meanings: (i) to aim at or shoot at, or to endeavour after; (ii) to endeavour to make out or guess anything, or to conjecture. Both seem to be relevant to modern usage.

In Plato's *Republic* we find *στοχάζεσθαι τοῦ μεγίστου ἀγαθοῦ*, "to endeavour after the greatest good"; while I was amused to find from Antipho (a Greek orator of about 440 B.C.) *ἄλλου στοχαζόμενος ἔτυχε τούτου*, which means "aiming at one thing, he hit another". I could not help thinking once more of J. B. S. Haldane's remark in the Karl Pearson memorial lecture that Pearson was like Columbus, who thinking he was going to China discovered America, but was not thought any the less of on that account.

However, I am getting too discursive. I must return to my most agreeable duty. I have much pleasure in seconding the vote of thanks to the President for his Address and congratulating him, on behalf of us all, on a most successful year of office.

Biomathematics

Reprinted by permission of
the Clarendon Press, Oxford

1. *Introductory remarks on biomathematics and biometry*

INAUGURAL lectures are usually somewhat predictable in their content, though I believe this University has had its exceptions. Some responsibility lies with the lecturer to explain to his audience what his subject is all about, and perhaps also to indicate what aspects he and his colleagues are particularly interested in. With familiar scientific subjects, like physics, chemistry, and biology, or even biophysics and biochemistry, the first responsibility is less important; but I think biomathematics sounds sufficiently novel to most people for me to have some obligation to dwell on it. I have made it the title of this lecture because it is the subject I have been appointed to profess; but I should remind you that my department has an evolutionary history, and developed from previous departments in Oxford, first under D. J. Finney and then under N. T. J. Bailey, with teaching and consultative commitments in biometry and statistics to the research schools and institutions of the University, especially in the biological sciences and medicine. The emphasis on biomathematics implied by the name of the present department is no doubt intended to reflect the rapid increase in recent years in the use of mathematics in biology (cf. also Bailey 1967), but it is qualified by my official duties, which include explicit reference to statistics. I would not want it otherwise, for I regard a proper balance of statistics and mathematics, or, more specifically, of biometry and theoretical biology, of the utmost importance in many current biological developments, so much so that I made the substance of a presidential address (1966) to the British Region of the Biometric Society a discussion of the relation of biometry with theoretical biology.

The gist of my thesis then was that biometry and biological mathematics are closely linked; but they should not be completely identified, or biometry include the whole of theoretical biology. The increasing use of mathematical models in biology makes this relation even closer, and more essential to maintain in a healthy state. As Professor E. J. Williams, of Melbourne, when discussing the development of biomathematical models, said (Williams 1967):

> The mathematical model of any phenomenon makes a contribution to understanding only if it leads to conclusions that can be checked against observational or experimental evidence. . . . Theories that do not pass this operational test are then little more than pure mathematics, even if the objects discussed are described in biological terminology.

The use of theoretical models is one that exasperates some biologists, familiar as they are with all the complexities of Nature. However, models could be interpreted so widely as to include all theories and hypotheses, although I personally prefer to use the word with some implication of approximation or idealization. As I said in my book on *Stochastic Population Models* (1960):

> The word 'model' is used deliberately, to imply the replacement of the complex biological reality by some more idealized hypothetical system.

Nevertheless, it is clear that there is no hard-and-fast division between a theory and a model. We hear the word 'model' much more often in biology or economics or operational research, where the need for some simplification is more evident, than in physics and chemistry; but in complicated physical situations physicists are not averse to using models in just the same sense, as in a simplified model for a heavy nucleus or for the molecular structure of a fluid. It is a subtle question how useful any simplified model is, in biology or elsewhere, but the outright refutation of all models on the facile grounds that they are not identical with Nature is equivalent to the refutation of all theory, without which nothing whatsoever would make any scientific sense for us.

Some biologists prefer to work from a different direction. Thus in the domain of ecology, C. S. Holling (1966, p. 5) has argued:

It would . . . be an entertaining and potentially profitable undertaking to analyse the various processes that affect animal numbers —e.g. predation, competition and parasitism—by initially ignoring the limitations imposed by the restrictions of traditional mathematical models and emphasizing the need for a realistic explanation. Once the analysis is complete it might then be possible to express the explanation in a precise mathematical form without sacrificing reality. In this way the process itself would dictate the form of the model rather than some arbitrarily chosen mathematical language.

There is no doubt that the advent of electronic computers has facilitated the introduction of more realistic assumptions into biological and other models. I am, however, a little sceptical of the returns from attempting too much in this direction. Speaking in my more empirical role as a statistician, I am conscious of the need in general to classify biological variability into the larger more dominant components and the innumerable smaller unimportant ones. Some compromise is always needed, or we reach a *reductio ad absurdum*, as Holling seems to realize later on (p. 68) in his monograph. He adds, in relation to the topic of attack by predators:

It is quite unnecessary to insist on explanations so rigorously realistic that they are ultimately based on, say, the kinetics of enzyme action or on electro-chemical events in the nervous system. Such insistence is a road with no end, for enzyme kinetics and neural action can in turn be explained by subordinate processes of biophysics and biochemistry. The striving for excessive realism can lead to a model as mysterious as the process modelled.

In Bellman's words (1957, preface), 'the Scientist, like the Pilgrim, must wind a straight and narrow path between the Pitfalls of Oversimplification and the Morass of Overcomplication'.

2. *History and scope of biomathematics*

The use of mathematics in any particular science is indeed

a rather critical function of that science's state of development. Theoretical and possibly quantitative principles certainly cannot be formulated until a coherent body of fact and empirical analysis has begun to emerge. Many instances of this could be mentioned, but it will be sufficient to cite Darwin's great synthesis of the evolutionary process in terms of the theory of natural selection, on which I will be commenting later.

The extent to which the development and behaviour of living organisms can ever be fully explained theoretically is itself a controversial question. For example, Professor Michael Polanyi has argued (1967, p. 55) that:

> The life process is essentially the development of a fertilized cell, as the result of information imparted by DNA. Transmission of this information is non-chemical and non-physical, and is the controlling factor in the life process. The description of a living system therefore transcends the chemical and physical laws which govern its atomic constituents.

By way of comment on this assertion, I would note that explanations of physical–chemical phenomena were first attempted in terms of mechanical (including electromagnetic) laws. The emergence and power of statistical concepts would force us, however, to include statistical theory when formulating our basic theoretical laws, and this to my mind is already introducing an hierarchical structure to our explanation, in which the statistical contribution is making an impact more at the level of what one might term 'system and organization theory'. At this level other specific logical formulations (as distinct from the general logical basis of all mathematical formulation) may be required, so that in the discussion of, say, error-free computers, or their living analogues, a combination of logical and statistical principles may be necessary. The irrelevance of equating such systems to any physical–chemical sets of components does not for me have any particularly profound significance. I suppose the reason is that mathematics is a more universal language than the narrower language of theoretical physics and chemistry. Not for one moment do I wish to imply by these remarks that we are all mere automata. Quite apart from the uncertainties that are part of the fabric of statistical laws, let us remember that the laws and concepts we are appealing to are themselves, at least in part, the products of our own imagination, so that we could hardly use them as an argument for complete determinism.

With regard to the *kind* of mathematics we can expect to require in biology, we have just seen that statistical and logical concepts are needed; and in my reference to statistical concepts I mean, of course, theoretical concepts rather than more empirical statistical methodology, important though this is, and in the development of which this country, and notably Karl Pearson and R. A. Fisher, played such an active role in the first half of this century. Having said that, and as these theoretical concepts are essential in *all* modern applied mathematics, I should stress that so far it is not the mathematics that is so different in biology from, say, in physics, but merely its area of application. It is true that control mechanisms and information theory, both comparatively recent developments, are very relevant, but so are they in, for example, operational research: and the mathematical theory of random or stochastic processes is universal in its application. It is particularly relevant in physics, in biology, and in operational research. Perhaps I might recall that when I returned to Cambridge after the war in 1946 and offered a lecture course on stochastic processes, I was asked by the mathematicians what this meant. Yet today (so much more sophisticated in our jargon have we all become) we find explicit reference to stochastic processes in such publications as Stafford Beer's *Decision and Control* and Carl Woese's *The Genetic Code*.

If, as against general classifications and mathematical tools, we look at individual fields of biological research where mathematics has played, or will play, an important part, we may conveniently divide these into one group involving the individual organism or its constituents and a second involving population structure and interaction. Among the first we may begin with the complete organism and its physical and morphological properties, pass on to various internal systems such as the metabolic and the nervous systems, and reach eventually the microscopic level of cells, chromosomes, genes, and DNA. At all these stages mathematical and statistical theory has already been introduced in the form of models, some more realistic and successful than others, of how the biological processes and functions operate. One of the most powerful and far-reaching is the genetic model initiated by Gregor Mendel, so that its limitations, and further explanation in terms of molecular biology, do not cause us to throw it out as mere 'bean-bag genetics', as Mayr (1963) has somewhat sarcastically referred to it, but where necessary to reinterpret it within a still wider and deeper class of phenomena.

The other main group of models, relating to populations, is one that, at least in some of its theoretical aspects, I am a little more familiar with. It would cover the fields of ecology, epidemiology, and population genetics, and I propose to survey the development and use of mathematical models in these fields in slightly more detail. My remarks will still no doubt appear highly superficial to those biologists who have made profound studies of particular areas, but I would plead that general surveys can have some use in helping us to see specific areas in their proper perspective, and perceiving more global relations that might be, and in fact have sometimes been, overlooked in investigations at too close quarters.

3. *Population models in ecology, epidemiology, and genetics*

Mathematical work in animal ecology largely dates from the publications of Lotka (1925) and Volterra (1926) on models of competition and predation (formulated technically in terms of a continuous time variable), and of Nicholson and Bailey (1935), who formulated their models perhaps somewhat less abstractly, in terms of discrete generations. However, in these early investigations little attempt was made to check in any detail the theoretical models with specific observational data, the reliable collection of which is always such a formidable task for the ecologist. A significant step forward was taken by Gause when he attempted to bridge the gap between theoretical models and actual populations by controlled laboratory experiments in animal ecology, particularly on prey–predator systems. These experiments he described in *The Struggle for Existence*, published in 1934. Various further investigations have been undertaken since then, including one very comprehensive series on competition, between two species of the flour-beetle *Tribolium*, by Park, in Chicago.

In epidemiology a parallel development was under way. Among early writers like Farr, Hamer, Brownlee, and Ronald Ross, the name of Ross is perhaps most familiar for his work on malaria; this included, though not before the second (1911) edition of his book on *The Prevention of Malaria*, a specific mathematical model. A succession of further studies appeared by Soper, Greenwood, Kermack and McKendrick, E. B. Wilson, and others. A close analogue of Gause's ecological experiments was a series of investigations undertaken by Greenwood and his co-workers and recorded in the Medical Research Council's *Report on Experimental Epidemiology* in 1936.

Most of the early theoretical models employed in epidemiology, as in ecology, were what we call *deterministic* in character, that is, they neglected all random fluctuations associated with the finite size of the populations. A feature of more recent discussion, for example, by myself and by Bailey (1957) in the field of epidemiology, is a more comprehensive formulation in terms of the theory of stochastic processes. Some of this work we later realized, however, had been anticipated by McKendrick, whose brilliant pioneering papers on stochastic models (1914, 1926) had been quite overlooked until their rediscovery by D. G. Kendall and J. O. Irwin (a historical review of epidemic theory by Serfling (1952) makes no reference to this work; for a more recent review, see Dietz (1967)).

There has perhaps been some doubt on the value of introducing further random elements into models that are already intractable enough; and it is certainly not my own intention to introduce complication unnecessarily. The important point to emphasize is that there are features of the stochastic formulation that are essential to our understanding of some crucial properties of finite biological populations, particularly in regard to the chances of extinction, whether these chances relate, say, to the onset of an epidemic, or to the preservation of a favourable genetic mutation. Remarkably enough, this extinction problem had already been raised, by Francis Galton, for example, in an appendix to *Natural Inheritance* (1889), in his discussion on the tendency of surnames to die out. It can even be linked with a very old problem in probability known as the problem of gamblers' ruin.

In the case of an epidemic arising from the transmission of infection by contact from infected person to susceptible person, it is known from the so-called Threshold Theorem in the deterministic formulations of Kermack and McKendrick that the susceptibles in a 'closed population', that is, constant in its total size, must reach a certain critical threshold in density before an epidemic will occur. The feature of the stochastic analogue is a richer and more realistic representation of the possible outcomes, as stochastic models deal not with certainty but with an entire distribution of probabilities. When the threshold density is exceeded, the probability distribution for epidemic size splits up and becomes bi-modal. Now when probability distributions are uni-modal it is often reasonable to summarize them by their mean, a simplification frequently made justifiable use of by physicists as well as by biologists; but it can hardly be recommended for bi-modal distributions (in spite of the familiar expression 'expectation of life', which is based on a bi-modal probability distribution of mortality by age).

My own work in epidemiology has included an extension of these ideas to a population with increasing susceptibles, as any natural human population is because of its children. The deterministic model, first formulated, for example for measles, by Hamer in 1906, predicts an oscillating tendency which we may, very roughly speaking and omitting some rather important complications, identify with the recurrent epidemics that we observe. Again, however, the stochastic formulation is more complete, and predicts a further threshold effect dependent on the size of the community; below a certain critical size the infection cannot maintain these oscillations but tends to become extinct. This phenomenon is well known for measles in extreme cases such as small island communities, but I demonstrated its relevance also for urban communities both in England and in the United States, the critical size of community being observed to be around a quarter of a million. A further observational check has more recently been reported by Black (1966). Below this critical size the natural epidemic periodicity is disrupted, and is dependent on new infection from outside, as I have shown for rather isolated communities as in Wales, where the normal natural period of around two years may lengthen considerably. Of course, this epidemic pattern for measles will soon be merely historical, in view of the introduction of the new vaccine.

Such a pattern will also not be maintained for another virus that can assume a more dormant or latent state. If an infected individual retains after recovery the virus ready for reactivation, no complete extinction of the virus will occur and the epidemic pattern will be modified. This is so for the chickenpox (varicella) virus, which appears to re-erupt in cases of shingles (herpes zoster). As for problems in ecology, detailed quantitative investigation is handicapped by the difficulty of estimating the relevant parameters—a crucial one in the chicken-pox situation is the ratio of chicken-pox cases arising from chicken-pox and herpes zoster respectively. These different possible forms of virus activity have important evolutionary implications, as noted both by Dr. Black and by Dr. Hope-Simpson (1965). The survival potential of the chicken-pox virus appears at a considerable advantage over that of measles, at least in the days of small primitive communities; this suggests that the measles virus is of more recent origin.

The theoretical possibility of local or temporary extinction underlines a general point which I do not think was sufficiently appreciated at one time, for example, either in connection with Gause's ecological experiments or Greenwood's epidemiological experiments. The crucial dependence of certain features on the *size* of the population implies, whether or not the relevant natural populations happen to be large enough for such dependence to be ignored, that experimental work on laboratory *populations* is especially sensitive to their size. Quite apart from other possible differences between laboratory and field conditions, the laboratory population may be too small to be directly comparable with its natural counterpart, and any results can then only be interpreted through the intermediary of a stochastic model.

If I turn next to the evolutionary theory of population genetics, it is with even more diffidence, and also with the sense of loss that the deaths in the last few years of its two great pioneers in this country, R. A. Fisher and J. B. S. Haldane, leave with us. This lecture has been a convenient goad for me to further my own acquaintance with this difficult and treacherous terrain, but I shall be relying largely, and no doubt sometimes rashly, on trails already blazed. That I refer to this topic at all is partly because of its general importance as a branch of biomathematics, and partly because the problems of ecology, epidemiology, and population genetics are in any case interrelated, both biologically and theoretically. Theoretically, because so many of the technical mathematical problems, in the general domain of stochastic process theory, are similar; biologically, because epidemiology is strictly a branch of ecology, and ecological problems cannot be fully discussed without genetical questions arising, such as the development of resistant strains of virus or bacteria. Conversely, evolutionary problems in genetics only make sense in relation to the ecological situation.

Mendel's genetic mechanism of inheritance had been rediscovered early enough in the present century to provide the theoretical framework on which Fisher, Haldane and, in the United States, Sewall Wright, were able to develop a quantitative theory of evolution by natural selection. Fisher, in his classic *The Genetical Theory of Natural Selection* (1930), underlined the essential role of genetic segregation in preserving, in the absence of natural selection and mutation, genotypic variability from one generation to the next. Natural selection could then operate at leisure on these different genotypes, with mutation merely providing the occasional but continual source of further genotypic variety on which selection could act. Both Fisher and Haldane, in their discussion on the temporal changes in the genotypic frequencies, usually assumed populations large enough for deterministic formulations to be adequate, apart of course from the important problem of the survival of mutant genes, to which I will return in a moment. The need to consider the effective finite size of populations has been more particularly emphasized by Sewall Wright in his so-called theory of random drift, which calls attention to the purely stochastic differences that may become fixed in different subgroups of the total population if these are sufficiently isolated. The biological relevance of Wright's thesis is still a matter of some controversy, the resolution of which has not always been assisted by the cloud of dust thrown up by the protagonists. There are at least two different aspects to be clarified: on the theoretical side, the rather difficult mathematics has not facilitated a very precise quantitative statement of what is to be expected; on the biological side, checking the causes of evolutionary changes observationally is obviously extremely difficult, and not always possible with sufficient accuracy to discriminate between rival explanations.

Bearing this and other outstanding problems in mind, I would not agree with the remark made, no doubt somewhat flippantly, by Lewontin (1965), that:

> By the end of 1932 Haldane, Fisher and Wright had said everything of truly fundamental importance about the theory of genetic change in populations and it is due mainly to man's infinite capacity to make more and more out of less and less, that the rest of us are not currently among the unemployed.

I have already emphasized the importance to biology of the theory of stochastic processes; and, with the growing need to be clear on what is theoretical deduction and what is empirical verification, a more intensive study of evolutionary theory in terms of stochastic process theory has been initiated by various workers, notable contributors including Kimura and Moran. Such work has led to the realization that the mathematical theory is even more complex than was at first always implied. Thus Fisher's original Fundamental Theorem of Natural Selection, connecting the increase in mean 'fitness' of a population through natural selection with the amount of genetic variability present, has been found to be rather restricted in its applicability. It has, for example, been shown by Moran (1964) that when more than one gene locus is involved, mean fitness, which is defined rather technically as an effective reproduction rate per generation averaged over the different genotypes, does not even necessarily increase. The further effect of linkage in the case of more than one locus is also extremely complex (see Ewens 1968).

We have seen that more specialized theorizing in biology has its dangers. In particular, Haldane would have been one of the first to criticize unnecessarily abstract or sterile discussion; at the same time, he often urged more mathematical effort in the study of evolution, and contrasted the enormous disparity in the time spent on mathematical problems in physics with those in biology (cf. Waddington 1968).

Much of the intricacy of the mathematical theory of evolution is associated with the rather cumbersome algebra of interbreeding genetic populations under selective forces. There is therefore something to be said for representing evolutionary processes where possible by approximating diffusion models, even at the risk of smoothing out some of the anomalies such as decreasing mean fitness. It would be even simpler for us if the difficulties of tracing the detailed evolutionary path, gene by gene, genotype by genotype, gene complex by gene complex, could be bypassed by appealing to more universal principles.

Such bypassing might be compared in physics with the appeal to general, for example thermodynamic, principles without an explicit analysis in terms of the relevant microscopic mechanisms. However, in spite of the great power of principles of this kind, they have in the absence of this detailed analysis an element of empiricism—consider, for example, the principle of least action—that can only be justified by sufficient predictive value which in turn stimulates and is verified by further observation. Today Darwin's unifying theory of evolution by natural selection is being linked with the discoveries of molecular biology and in particular with the universal nature of the genetic code, so that there is no longer any doubt of the rapid trend towards a biological synthesis that could hardly have been dreamed of even a generation or so ago. Yet it is, I think, fair to say that a corresponding *quantitative* synthesis has only just begun, and successful general principles in biology embodying *quantitative* prediction are still very much in their infancy.

In the Biometric Society address to which I referred earlier I mentioned one or two tentative and not altogether convincing principles of this kind, including a Principle of Optimum Design put forward by Rashevsky, and an optimum principle proposed by Kimura (1958) in the theory of natural selection intended to play a role analogous to that of least action in physics. However, in more recent years Kimura (1960) has taken up a concept originating with Haldane, that of genetic load; and has proposed what he terms the principle of *minimum* genetic load. The evolutionary relevance of genetic load, which is measured in terms of the difference between the mean fitness of a population and the optimum fitness, is not immediately obvious. A slight digression at this point on the problem of the survival of mutant genes may be helpful.

We know that a population's survival depends on whether its effective reproductive rate is greater or less than unity. This leads to a paradoxical instability, for either the population dies out or increases without limit. It will not even do to have its reproductive rate precisely equal to one, for stochastically this still leads to eventual extinction for any finite population. This strictly leads us to the more complicated control mechanisms of ecological density-dependence, where the reproductive rate adjusts to the population density (cf. Turner and Williamson 1968); but it is hardly yet possible to introduce such complications very satisfactorily into theoretical discussions of evolution, and it is more usual to assume that if a population is roughly in equilibrium it must have its reproductive rate about one. The fate of a new mutant that increases fitness is still somewhat precarious, for it is the absolute, not the relative, fitness that is relevant, and only if this absolute fitness is greater than unity does the new mutant have a chance to survive. This argument applies directly to selfing haploid individuals, but may be applied also to the reproduction of sexually differentiated diploid individuals if the fitness of the mutant gene refers to the fitness of the heterozygote, which alone will influence the survival of the gene when it first appears. Curiously enough—at least, I have not noticed the point made in the literature—it does not apply without modification to selfing diploids. In this case, both genotypes involving the mutant gene have to be considered. If the absolute fitness of the original homozygote is 1, and of the new heterozygote $1+s$, and of the mutant homozygote $1+s'$, the chance of survival is zero unless $\max\{\frac{1}{2}(1+s), 1+s'\}$ is greater than one. Even when the homozygous mutant has the same fitness as the heterozygote, the chance of survival is for small s only half its value for the usual sexual reproduction case. The biological advantages of sexual reproduction are usually envisaged in terms of the greater availability of the existing genetic store of material, but greater chances of survival of new advantageous mutants would also seem of some relevance. Incidentally, these advantages all relate to the contrast of selfing with crossing rather than with sexual differentiation as such, so that the further evolution of sexual differentiation of individuals would not be explicable purely on these grounds.

To return to the concept of genetic load, the gist of the argument that Haldane put forward, as I interpret it, was that if the optimum fitness could not be expected to differ much from unity the mean fitness could not have differed excessively from it for any length of time, or the population would have already become extinct. A constraint is consequently imposed on the nature and speed of possible evolutionary change. The argument is hardly precise enough for acceptance as other than suggesting a principle to be studied and tested; as such, its value must depend on the extent of its predictive power and success.

There are at least three sources of genetic load: (i) mutational, (ii) segregational, in the case of stable polymorphisms, and (iii) substitutional or evolutionary, due to the replacement of a gene by a more successful allele. In the case of a haploid organism, the substitutional load l_e, measured (following

Kimura) in terms of its contribution to the logarithmic growth-rate of the population, is simply

$$l_e = -\log_e p \tag{1}$$

per unit time (in generations) per gene substitution, where p is the initial relative proportion of the gene that is taking over. In the more usual diploid case this formula is replaced in general by a more elaborate one,

$$l_e = -\frac{1}{h}\left\{\log_e p + (1-h)\log_e \frac{1-h}{h+(1-2h)p}\right\}, \tag{2}$$

where h is the degree of dominance of the new allele over the original wild type. This result is derived on the basis of an initial equilibrium value of p given by

$$\mu = sp\{h+(1-2h)p\}, \tag{3}$$

where μ is the mutation rate and s the original *decrease* in fitness of the mutant homozygote before a changed environment gives it a net *increase* in fitness s' over the wild type. The mutational load in the original environment may be written

$$l_m = sp\{p+2h(1-p)\}. \tag{4}$$

The minimization of the quantity

$$L_m + L_e = \sum (l_m + \epsilon l_e), \tag{5}$$

where the summation is over all gene loci (treated as independent), and ϵ is the rate of substitution per locus and is dependent on the rate of environmental changes, led to various predictions that hinged on the values of the two quantities

$$D = \sum sp, \qquad E = \sum \epsilon. \tag{6}$$

The quantity E had already been estimated by Haldane (1957) from the observed speed of evolution of phenotypic characters to be about 1/300. The quantity D was taken by Kimura to be 2, using some data on inbreeding depression in fitness for man, from a paper by Morton, Crow, and Muller (1956).

With these values the predicted value of $\bar{h}$, the harmonic mean of h, worked out at 2 per cent, which seemed to agree quite well with an estimate available from *Drosophila*. The predicted value of the total mutation rate $\sum \mu$ was 6 per cent, compared with an estimate, based on some results of Muller's, for both man and *Drosophila*, of the order of 10 per cent. The value of $\bar{s}$, the harmonic mean of s, was 1 per cent, with the arithmetic mean rather higher at about 2 per cent. The total mutational and evolutionary loads were about 10 and 20 per cent respectively.

While the predicted quantities in these calculations obviously depend both on the validity of the theory and on the values of D and E inserted, it should be noticed that not only does this theory link the optimum mutation rates and degree of dominance, themselves subject to selective modification, with other quantities such as selection coefficients, but does so quantitatively. I should stress again, with Kimura, that the evolutionary relevance of genetic load, and in particular the numerical implications of his principle, would be a matter for justifiable caution, indeed even scepticism among many population geneticists (cf. also, for example, Kimura 1967); but some at least of the criticisms of this concept appear to arise from misunderstanding. For example, Dobzhansky (1965) has said:

The concealed genetic loads are, however, far from alike in all populations. And contrary to what one might have expected, they are neither the lightest in the healthiest and most successful nor the heaviest in least prosperous populations. Thus, *Drosophila willistoni*, the ecologically versatile and extremely common species in much of tropical America, has the heaviest, while *D. prosaltans*, ecologically apparently specialized and usually rare species, has the lightest load. . . . In *D. pseudo obscura*, a species commonest in the temperate zone of western N. America, the lightest load is found in the geographically outlying population isolated on the Andean plateau, near Bogota, Columbia.

But the arguments to which I have just referred relate the mutational load to be expected directly with the environmental changes, and appear to me to be not inconsistent with Dobzhansky's observations, at least (as we have not yet considered *segregational* load) if these indicate variation in *muta-*

tional load. The total predicted mutational load of 10 per cent (which only partially agreed with the evidence from *Drosophila*) was in the nature of an average, and would be expected to increase with the heterogeneity of the environment.

An examination by Kimura of the segregational load, which might appear at first sight purely deleterious (it is hardly sufficient to say that the heterozygote is more fit than the homozygotes, for modification by selection could alter this), confirmed its relevance in, for example, alternating environments, the theoretical optimum fitness being then equally poised with its maximum at the heterozygote. The variation of the genotypes in fitness is, however, still a contribution to the genetic load, which could be diminished by a reduction in such variability, i.e. by an evolutionary improvement in the homeostatic properties of the organism.

In the case of heterozygosis involving more than one gene locus, the segregational load is also a function of the linkage between the two or more loci, and it may be shown that the load is reduced by closer linkage (see, for example, Ewens 1968).

These calculations on minimum loads have so far neglected the finite size of populations, but from my earlier remarks we shall not be surprised if for some purposes this is not adequate. In the case of polymorphisms due to heterozygous advantage at a gene locus, the segregational load would be reduced by increasing the number of alleles, but the finiteness of populations imposes a limit on this. Some calculations by Kimura and Crow (1964) suggest that for a selective superiority of 1 per cent in all the heterozygotes and a mutation rate of 10^{-5} at a single locus, a population of size 10 000 can maintain about eight alleles at this locus, with a corresponding contribution to the segregational load of about 10^{-3}.

The finite size of populations has been emphasized more recently by Kimura in an estimate of evolutionary rates at the molecular level. In an article in *Nature* (1968), he gives a formula for substitutional load depending on effective population size, and, incidentally, also on the selective advantage of the new mutant leading to its probable fixation (the earlier formula (2), effectively due to Haldane, does not depend on this selection coefficient). The mathematical derivation of the new formula is not given, but appears to be based on the diffusion approximation model for a finite population, this being most applicable for small values of the selection coefficient. The formula is used in connection with a rather surprising estimate of nucleotide substitution at the rate of one every two years, compared with Haldane's estimate based on the evolutionary change of phenotypic characters of one every 300 generations. This comparatively rapid rate is calculated from an estimated amino-acid substitution in a polypeptide chain, of length 100 amino-acids, of one in 28×10^6 years, coupled with the figure of 4×10^9 nucleotide pairs in the (haploid) chromosomal set for man. The substitutional load implied on the earlier formula would appear intolerable, but it is possible on the amended formula if the mutations are practically neutral in effect. Kimura suggests that this apparent plethora of neutral mutations at the molecular level may greatly increase the role of random drift in evolutionary change.

The relation of genetic structure with the structure of DNA extends to their assessment in terms of information theory. In contrast with the mutational load, which is the price paid for maintaining the appropriate flow of mutations and as such provides no net gain in information, the substitutional load represents a real gain, which from formula (1) is directly interpretable as an information quantity. More generally, this is $L_e/\log_e 2$ in 'bits' per generation. The estimate by Kimura in 1960 of the total information accumulated over the last 500 million years or so up to man was 10^8 bits, based on the value 20 per cent for L_e. This compared with an estimate of the potential information on the (diploid) chromosomal set for man of the order of 10^{10} bits. We should be wary of comparing figures of this kind, obtained from very different considerations; the figure 10^8 is the estimated information measure of the gain in the survival probability, whereas the figure 10^{10} is the estimated potential information capacity of the nucleotide sequence. However, any genetic information must be contained in the DNA structure, and the comparative values are at least consistent, even the difference being ascribable partly to redundancy in the coding mechanism. The estimate 10^8 does not, moreover, appear to need modification to take account of the much more rapid nucleotide substitutions, for these, if neutral, do not add to the genetic information, at any rate that associated with survival. The amount of redundancy in the coding mechanism also appears to change systematically as organisms evolve to higher forms; though I do not understand a recent discussion of this topic by Gatlin (1966, 1968), as not only does he not take proper account of the triplet code structure of the nucleotide sequence, but appears to identify redundancy with the total information content, instead of with its partial loss but corresponding compensation by lower probabilities of transcription errors.

The switch from the concept of fitness via genetic load to that of information content I, say, has an appeal as a measure of evolutionary complexity, and raises the question whether some over-all measure or index that takes proper note of evolutionary success is possible. My own provisional and admittedly arbitrary suggestion for a homogeneous population for which internal interactions or genetic variability among its individual members are unimportant is

$$S = NI, \tag{7}$$

where N is the present effective population size and I the genetic information content in the individual organism. More precisely, I would refer to the genetic information stored in the germ plasm, this including not only the component that has evolved from past variation in fitness but any additional component (excluding, of course, redundancy), for example, any attributable to random drift. For a non-homogeneous population this could be generalized to

$$S = \sum_{r=1}^{N} I_r; \tag{8}$$

and such a formula could even be summed over different populations, at least where non-interacting.

A complete evolutionary theory should account not only for degree of dominance and optimum value of mutation rates, but also for sexual differentiation, species differentiation, evolution of the chromosomal apparatus, and the genetic code. With regard to sexual differentiation, I pointed out earlier that crossing is to be distinguished from sexual differentiation, and Geodakyan (1965) has remarked that a population of two differentiated sexes is in some ways at a biological disadvantage compared with a population of monoecious or hermaphrodite but outbreeding individuals. He argues that in a sexually differentiated population the *number* of offspring is mainly determined by the female population and the *quality* by the males, on the grounds that with more males the potentialities for selection are improved. His discussion is rather qualitative, but contains some interesting comments on the optimum sex ratio and the ratio of male to female variability. One might summarize his discussion somewhat loosely in terms of my over-all index S by allocating the males a greater role in increasing I and the females a greater role in increasing N, though this suggestion is necessarily dependent on what biological significance we may be prepared to ascribe to the single quantity S as a measure of evolutionary success.

As a *caveat* to my temerity in even considering such a quantity, let me remind you that the taxonomists have been faced with the intricate task of classifying the millions of species that exist today. Some of them are beginning to use quantitative methods, and some of these in turn (in the absence of detailed knowledge of information differences in the DNA sequences) are formulated in terms of phenotypic 'distances' between species also based on information concepts; but this no more eliminates the formidable task of studying all the intricacies of these interrelations (whether at the phenotypic or the molecular level) than the astronomers in possession of one or two general physical principles are absolved from studying all the awe-inspiring wonders of the heavens.

The effect of interactions, not only between individuals, but also between populations, certainly needs much further consideration. Thus the problem of density-dependence which I have already mentioned is even more complicated when it involves two or more interacting species. For example, in a typical prey–predator relation, the reproductive capacity of the prey will only genetically increase at this population's peril. If it does, as noted by MacArthur (1965) and indeed by Nicholson and Bailey (1935), it is the predators that tend to increase at the expense of the prey. This raises combined evolutionary–ecological problems at a still higher hierarchical level of complexity, which (quite apart from the effect on any measure such as S) have hardly yet been satisfactorily formulated.

Finally, let me come back to where I started, on the place of mathematics in biology. I cannot do better than end by quoting from Kostitzin's *Mathematical Biology* (see § 2), published about thirty years ago:

> Everyone is agreed that figures have the right of entry into Natural Science, but when it becomes a matter of reasoning about these figures, or of dealing mathematically with them, there is encountered a resistance, a repugnance. Why is this ? It must be admitted that this resistance does not always come from narrow-mindedness or from conservatism. Reasoning in general does not frighten a naturalist, but mathematical reasoning startles him, because he is in the habit of verifying each step by experiment.

In reasoning, experimental results are subjected to a series of logical operations. The accuracy of the ultimate result depends on that of the initial data, and also on the number and nature of the logical operations carried out between the premise and the conclusion. In ordinary reasoning, this number is not very great and successive stages are always verifiable. In mathematical reasoning, the steps are taken too quickly and a result is reached which may appear to be, and very often is, arbitrary or untrue. The reason for this is that in establishing a biological equation, the problem is simplified by sacrificing a number of factors or a number of details, and these sacrifices distort the results that emerge from the logical method. Inexact yet acceptable initial hypotheses are transformed into gross errors by a few turns of the crank of the logical apparatus. Simplification gives rise to paradox. This is a fact that cannot be denied, and one that is common to all experimental sciences. In purely mathematical problems, the mathematician is certain of his premise and the number and nature of the operations is of interest only from the aesthetic point of view; but in dealing with applications of mathematics, the position is quite different. A long period of development has been necessary to secure unquestioning acceptance of the agreement between reasoning and experiment in mechanics, physics and astronomy. Perhaps in biology this development will be more rapid by virtue of an acquired impetus. In any case, long and continuous collaboration between mathematicians and biologists is necessary before reaching the same assurance as in the physical sciences.

References

(Some of these are not explicitly referred to in the text)

BAILEY, N. T. J. (1957), *The mathematical theory of epidemics,* London.

——— (1967), *The mathematical approach to biology and medicine*, John Wiley, New York.

BARTLETT, M. S. (1960), *Stochastic population models in ecology and epidemiology*, Chapman and Hall, London.

——— (1966*a*), *Introduction to stochastic processes*, 2nd edn, Cambridge University Press.

——— (1966*b*), Biometry and theoretical biology. Presidential address to British Region of Biometric Society.

BEER, S. (1967). *Decision and control*, John Wiley, New York.

BELLMAN, R. (1957), *Dynamic programming,* Princeton University Press, New York.

BLACK, F. L. (1966), Measles endemicity in insular populations: critical community size and its evolutionary implication. *J. theor. Biol.* **11**, 207–11.

CROW, J. F., and KIMURA, M. (1965), The theory of genetic loads. *Genetics today*, **3**, 495–506.

DIETZ, K. (1967), Epidemics and rumours: a survey. *Jl R. statist. Soc.* **A130**, 505–28.

DOBZHANSKY, T. (1965), Genetic diversity and fitness. *Genetics today*, **3**, 541–52.

EWENS, W. J. (1968), *Population genetics*, Chapman and Hall, London.

FELLER, W. (1967), On fitness and the cost of natural selection. *Genet. Res.* **9**, 1–15.

FISHER, R. A. (1930), *The genetical theory of natural selection,* Chapman and Hall, Oxford.

FORD, E. B. (1964), *Ecological genetics*, Chapman and Hall, London.

GALTON, F. (1889), *Natural inheritance,* London.

GATLIN, L. L. (1966), The information content of DNA. *J. theor. Biol.* **10**, 281–300.

——— (1968), The information content of DNA II. *J. theor. Biol.* **18**, 181–94.

GAUSE, G. F. (1934), *The struggle for existence*, Hafner, Baltimore.

GEODAKYAN, V. A. (1965), Role of the sexes in the transmission and transformation of genetic information. *Problemy Pered. Inf.* **1**, 105–12.

GREENWOOD, M. *et al.* (1936), *Experimental epidemiology*, M.R.C. Special Report 209.

HALDANE, J. B. S. (1930), A mathematical theory of natural and artificial selection. *Proc. Camb. Phil. Soc. math. phys. Sci.* **27**, 137–42.

——— (1937), The effect of variation on fitness. *Am. Nat.* **71**, 337–49.

———(1957), The cost of natural selection. *J. Genet.* **55**, 511–24.

HOPE-SIMPSON, R. E. (1965), The nature of herpes zoster: a long-term study and a new hypothesis. *Proc. R. Soc Med.* **58**, 9–20.

KIMURA, M. (1958), On the change of population fitness by natural selection. *Heredity*, **12**, 145–67.

——— (1960), Optimum mutation rate and degree of dominance as determined by the principle of minimum genetic load. *J. Genet.* **57**, 21–34.

——— (1961), A measurement of the rate of accumulation of genetic information in adaptive evolution. *Bull. Inst. int. Statist.* **38** (3) 239–48.

——— (1964), Diffusion models in population genetics. *J. appl. Prob.* **1**, 177–232.

——— (1967), On the evolutionary adjustment of spontaneous mutation rates. *Genet. Res.* **9**, 23–34.

——— (1968), Evolutionary rate at the molecular level. *Nature, Lond.* **217**, 624–6.

——— and Crow, J. F. (1964), The number of alleles that can be maintained in a finite population. *Genetics*, **49**, 727–38.

KOSTITZIN, V. A. (1939), *Mathematical biology*, London.

LEWONTIN, R. C. (1965), The role of linkage in natural selection. *Genetics today*, **3**, 517–23.

LOTKA, A. J. (1925), *Elements of physical biology*, Dover, Baltimore.

MACARTHUR, R. H. (1965), Ecological consequences of natural selection, in *Theoretical and mathematical biology*, New York.

MCKENDRICK, A. G. (1914), Studies on the theory of continuous probabilities with special reference to its bearing on natural phenomena of a progressive nature. *Proc. Lond. math. Soc.* (2) **13**, 401.

——— (1926), Applications of mathematics to medical problems. *Proc. Edinb. math. Soc.* **44**, 98–130.

MAYR, E. (1963), *Animal species and evolution*, Hutchison, Cambridge, Mass.

MORAN, P. A. P. (1962), *The statistical processes of evolutionary theory*, Oxford University Press.

——— (1964), On the nonexistence of adaptive topographies. *Ann. hum. Genet.* **27**, 388–93.

——— (1967), Unsolved problems in evolutionary theory. *Proceedings of the 5th Berkeley Symposium on Mathematical Statistics and Probabilities.*

MORTON N. E. (1965), Models and evidence in human population genetics. *Genetics today,* **3**, 935–51.

MORTON, N. E., CROW, J. F., and MULLER, H. J. (1956), An estimate of the mutational damage in man from data on consanguineous marriages. *Proc. natn. Acad. Sci. U.S.A.* **42**, 855–63.

NICHOLSON, A. J., and BAILEY, V. A. (1935), The balance of animal populations. *Proc. zool. Soc. Lond.* 551–98.

POLANYI, M. (1967), Life transcending physics and chemistry. *Chem. Engng News* 21 August, 54–66.

ROSEN, R. (1967), *Optimality principles in biology*, Butterworths, London.

ROSS, R. (1911). *The prevention of malaria*, 2nd edn., London.

SERFLING, R. E. (1952), Historical review of epidemic theory. *Hum. Biol.* **24**, 145–66.

SOKAL, R. R., and SNEATH, P. H. A. (1963), *Principles of numerical taxonomy,* W. H. Freeman, New York.

STAHL, W. R. (1967), The role of models in theoretical biology. *Prog. theor. Biol.* **1**, 166–218.

TURNER, J. R. G. (1967), Mean fitness and the equilibria in multilocus polymorphisms. *Proc. R. Soc.* **B169**, 31–58.

——— and WILLIAMSON, M. H. (1968). Population size, natural selection and the genetic load. *Nature, Lond.* **218**, 700.

VOLTERRA, V. (1926), Variazionie fluttuazioni del numero d'individui in specie animali conviventi. *Memorie Accad. pont. Nuovo Lincei,* **2**, 31.

WADDINGTON, C. H. (1968), Towards a theoretical biology. *Nature, Lond.* **218**, 525–7.

WILLIAMS, E. J. (1967), The development of biomathematical models. 36th session, International Statistical Institute, Sydney.

WOESE, C. R. (1967), *The genetic code*, Har-Row, New York.

WRIGHT, S. (1931), Evolution in Mendelian populations. *Genetics,* **16**, 97–159.

When is inference *statistical* inference?

Reprinted from *Foundations of Statistical Inference*,
a symposium, by permission of Holt,
Rinehart and Winston of Canada Ltd

Summary

The thesis is argued that a statistician qua statistician should either confine his inferences to statistical inferences in the sense indicated in this paper, or at least make it clear when his inferences are not so classifiable.

The arguments for and against prior probabilities, likelihood ratios, and alternative methods of estimation are discussed and illustrated in this context.

Introductory remarks

I have become more pessimistic over the years about the possibility of complete agreement over the principles of *statistical* inference. The difficulties are partly terminological and semantic, as exemplified by differences of opinion of the meaning and range of the words *statistics* and *statistical*; but over and above this there is the almost inevitable controversy over the nature and methods of all inductive inference. Statistical inference is certainly at least part of all inference; hence the wider controversy muddies the waters of the narrower stream. Statisticians, and others interested in the statistical inference problem, often seem to me to have refused to recognize the open-endedness of the problem by adopting one of two attitudes: either (a) they may ignore or belittle the inductive inference aspect of the statistical problem, when they discuss statistical *hypotheses* as if one of them at least were true (we can more realistically treat all those so far enunciated as false), or (b) (like H. Jeffreys and L. J. Savage) they are so attracted by the apparent unity and comprehensiveness of the Bayesian approach to all induction that they do not always recognize, or admit, that statisticians have a field of discourse that is not identical with the universe of all discourse (except that, curiously enough, their examples are often so dominated by statistical examples that I have claimed that they use the more robust statistical component of these problems to bolster up the weaker non-statistical part). One could classify these two approaches as (a) partial precision by unjustifiable segregation (b) spurious over-all precision created by analogy. My pessimism over the existence of a final complete solution explains my reluctance to add further to the spate of words that have been already written on this subject, a spate that should have warned us of the possible absence of any unique acceptable solution. I have suggested above that we have all been in danger at times of evasion and omission, and if this current round of discussion is to be more than a merry-go-round it could be because the participants are prepared to face difficulties frankly and impartially rather than indulge in what has been at best a debate between protagonists and at worst a slanging match.

Some of the omission stems from the historical accidents of our education and obligations, so that mathematicians discussing Bayesian theory often concentrate on the mathematics without bothering to ask themselves what they are talking about, statisticians undertaking statistical analyses for a scientific paper or for a client cannot confuse the immediate task by a digression on philosophical fundamentals. This limitation of their horizon, whether deliberate or not, is often quite right and proper. It may, as I have remarked before, be compared with the routine work of the physicist or biologist, who would merely bore his audience if he dragged the whole question of scientific method into his work at every conceivable opportunity. It is not a valid criticism of his work to say he has not discussed its foundations; it would only be a valid criticism if it were established that he had never considered these. Of course, to compare the statistician with the physicist or biologist is to claim that there are recognizable phenomena that may be classified as *statistical*, just as there are phenomena classified as physical and phenomena classified as biological. Such classification is inevitably bedevilled both by 'woolly' and overlapping frontiers, and by the philosophical conundrum of separating fact and theory; and in the case of statistics it has not apparently been accepted by everybody. However, like Hacking (1965), Edwards (1969) and others, I propose to accept the existence of sex-ratios, expectations of life, genetic re-combination fractions and the like, as much as velocitics, electrons, living organisms and cells. This necessitates the use of statistical *probabilities* if we are to have any kind of mathematical theory, and a theory of statistical inference which is not purely deductive, I would define as concerned with inference about such probabilities.

This attitude raises various questions and points which need enunciation and discussion. First of all, it is of course often necessary in practice to draw rather vague inferences about statistical quantities, just as about other quantities, because of the vague nature of the data–an example would be a rough prediction in epidemiology, or meteorology because of the heterogeneous or impure character of the data. The next point to raise is whether in appropriate more homogeneous or 'pure' statistical situations it is possible to employ specific techniques which would not be available elsewhere; and, if so, whether they are intrinsically statistical in their interpretation or can be applied to non-statistical questions, such as what will happen at a unique trial? There is an obvious and major difficulty here which affects my own general attitude to these questions. This is how we recognize the appropriate 'pure' statistical situation. To someone who is not prepared to consider such a situation, the question does not arise, but if phenomena are not to be classified at all, life is going to be hard indeed. However, I must remind you of my *caveat*, that in agreeing to isolate the statistical problem we must not ignore its general background. This is why I have never claimed that statistical methods or inferences are more than a partial answer to some of our real questions, and we must never forget this (cf. Skellam, 1969). It seems to me that my attitude is nearer to R. A. Fisher's earlier outlook, when he emphasized the reduction of data, the sampling properties of estimates, etc. than his later attitude, which was closer to that of writers like G. Barnard, Edwards and Hacking, when they discuss artificial examples, which seem at times to be in danger of being over-academic and narrow. My own reaction is still to formulate questions statistically in statistical situations while recognizing the limitations of some of the answers, and to regard an answer to a non-statistical question in a statistical situation as no more than a convention (which may be interesting, but no more necessarily logically acceptable than any other inductive inference). Questions and answers in non-statistical situations I would regard as non-statistical by definition, and hence outside my terms of reference.

This, I think, gives me a certain consistency of approach associated with a refusal to mislead by spurious over-all precision, and a realisation that we do not all seem to see eye to eye on these matters. As this statement of my own attitude is possibly unclear without illustration, I propose to add remarks on some of the more specific controversial techniques with a view to re-iterating my own position. Any apologies for repetition are qualified by the comment that others demonstrably do not always bother to read, or at any rate to take in, what has been previously stated, so that some degree of redundancy in one's own writings seems not only more justifiable, but almost essential for adequate communication.

Prior probabilities

In a review of the interesting little book by Maynard Smith on *The Mathematics of Biology*, I objected to one example and its solution, both seeming to me incompatible with the interpretation of probability as relative frequency given in the text of the book. I quote below the relevant portion of my review (1969a):

'The last example on p. 70 has nothing to do with probability as frequency, and requires detailed comment. The author says it nearly wrecked a conference on theoretical biology, but also says that it yields at once to common sense or to Bayes' theorem. If it is so simple, this does not seem to say much for the conference participants; moreover, as a question on Bayes' theorem, the answer given may legitimately be questioned. The example is as follows:

Of three prisoners, Matthew, Mark and Luke, two are to be executed, but Matthew does not know which. He therefore asks the jailer, 'Since either Mark or Luke are certainly going to be executed, you will give me no information about my own chances if you give me the name of one man, either Mark or Luke, who is going to be executed.' Accepting this argument, the jailer truthfully replied 'Mark will be executed'. Thereupon, Matthew felt happier, because before the jailor replied, his own chances of execution were 2/3, but afterwards there are only two people, himself and Luke, who could be the one not to be executed, and so his chance of execution is only 1/2.

Is Matthew right to feel happier?

The individuals to be executed are already determined and known to the jailer, who is therefore justified in passing on as irrelevant to Matthew the name of either Mark or Luke as one of these. However, it is said that Matthew is prepared in the absence of this certain knowledge to allocate numerical probabilities to possible eventualities as indicated. Is he being consistent in his assessments? Suppose the prior probabilities are denoted in general by p, q and r respectively for Matthew's, Mark's or Luke's survival. Then the posterior probability of his own survival after the jailer's information is from Bayes' theorem changed from p to $p' = pP/(pP + r)$, where P is the probability of the jailer naming Mark, if both Mark and Luke are to be executed. Note that *to the jailer* either p or r is 1, and in either case (whatever P) there is for him no change in p. Now *if* Matthew assigned values $p = q = r = 1/3$, $P = 1/2$, then also for him $p' = p$, and he was behaving inconsistently in claiming $p' = 1/2$. But he is quite justified (consistently with the wording of the example) in first assigning $p = 1/3$, $p' = 1/2$, provided he then chooses $r/P = 1/3$. His assessments are now all consistent, and, as $p' > p$, he has a perfect right to feel happier!

If the example proves anything, it is the indefiniteness of subjective prior probabilities.'

Professor Lindley, one of the most whole-hearted Bayesians in England, objected to my comments, and made the following reply (1969):

'In the course of a book review (*Nature*, **221**, 291; 1969) Professor M. S. Bartlett discusses the problem of three prisoners, two of whom are to be executed. His conments merit further discussion. He first remarks that if, as has been reported, the problem nearly wrecked a conference on theoretical biology and yet yields at once to Bayes's theorem, it does not say much for the conference participants. This seems a little hard on the theoretical biologists who will typically have learnt their probability from a member of the frequentist school who, if he mentioned Bayes's theorem at all, will have played it down as of minor interest. The fault surely lies with the statistician, not the biologist.

The second point is more material. Bartlett draws the conclusion that subjective probabilities are indefinite. The basis for this assertion seems to be that the prisoner, Matthew, is entitled to feel happier if P (the probability of the jailer naming Mark if both Mark and Luke are to be executed) equals $3r$. If $r > 1/3$, the Bayesion analysis shows that this is impossible so that Matthew is inconsistent. If $r = 1/3$, this requires $P = 1$. Taking $r = 1/3$ as a reasonable value, this shows that Matthew's elation is only justifiable (to him!) if he believes that when there is a choice, the jailer will always name Mark. So I would say to Matthew, "If you feel elated it is equivalent to your assuming this preference on the jailer's part". Matthew would typically reply that he has no reason for thinking the jailer has such a preference: therefore, I say, he has no reason for feeling happier. The subjective probabilities must cohere and their value lies in doing just this–in the example in establishing coherence between Matthew's happiness and his opinion of the jailer. This seems to me to lead to a definite conclusion of some value, contrary to what has been said.'

One or two other writers joined in, and I also had a personal communication from Harold Jeffreys, but none of these appear to me to have thrown any new light on the problem, and I do not propose to quote them all here. I merely add my own published reply (1969b) to Lindley:

'Professor Lindley (*Nature*, **221**, 594; 1969) is, like Matthew, entitled to his opinion, but I do not see that his comments in any way dispose of my criticisms. First of all perhaps I could protest at the irrelevant remarks in his first paragraph, which includes an inaccurate reporting of what I quoted Professor Maynard Smith as saying, namely, that the problem "yields at once to common sense or to Bayes's theorem". By omitting the phrase "to common sense or" Professor Lindley distorts, quite unfairly, the implication in my own remarks.

To come now to the problem, this is unsatisfactory partly because of the third-person reporting in it of Matthew's opinions as data to be used in any numerical evaluations. These data included Matthew's assessments of his own probability of execution as changing from 2/3 to 1/2, so that he had a right to feel happier in particular if it could be shown that he was being consistent. Professors Maynard Smith and Lindley apparently think he had no such right, but my claim that this view 'may legitimately be questioned' was illustrated by my demonstrating how Matthew could be consistent in his assessments. Lindley introduces the further assumption that $r = 1/3$ "as a reasonable value", and concludes that P, the probability of the jailer naming Mark if both Mark and Luke are to be executed, would then have to be 1. Dismissing this possibility, Lindley rejects the consistency of Matthew's assessments. However, even if I accept Lindley's value $r = 1/3$ (which Matthew may not have done). Matthew has still a 'right to feel happier' on the weaker basis that his second assessment should be at any rate $< 2/3$, which leads to the condition $P > 1/2$. I see nothing extraordinary in such a belief by Matthew about P, especially if he recalls the order of precedence of Mark and Luke in the New Testament (the jailer may not be a Christian, but Matthew may be unaware of this).'

Of course I am not arguing against the explicit use of prior probabilities when they are well-determined in the statistical sense, such as with the prior probability of like-sex twins being monozygotic, or in other empirical uses of Bayes' theorem as advocated by H. Robbins and other writers.

Likelihood ratios

If, like Hacking and others, we concentrate on the likelihood ratio as the relevant sample quantity to discriminate between two hypotheses (note that its relevance and sufficiency may be deduced from Bayes' theorem or directly, as was first done, I think, by B. L. Welch), we have yet to say how it is to be used. I agree with Hacking that without Bayes' theorem we have to decide what direction gives the greater plausibility to one as against the other alternative, but the direction (which would follow in any event from cases where Bayes' theorem has a frequency interpretation) is not one that I have ever heard queried, so it is not an unreasonable axion. Hacking suggested in his book that the axiom does not follow from long-run frequency – his counter-example* was not, however, too convincing, for he appealed to a minimax solution which made the 'long-run' solution obviously absurd. I discussed quantitative aspects of the long-run solution further in my review (1966), but agreed that this further discussion still did not itself justify the universally accepted solution to his problem. The more relevant comment appears to me to be supplied by Hacking himself, that someone 'may say that it is habit that makes you suppose that if there were to be no long run, there would still be a reasonable guess'. I do not object to the axiom that the hypothesis with the greater likelihood has the greater *support*, indeed, this is merely a definition of support, but different people have different reasons for their acceptance of such an axiom, and I do not see that either a Bayesian background of uniform prior probabilities, or a statistical background of long-run assertions based on likelihood ratios, can be dismissed as irrelevant if an individual chooses to make use of one of them. Indeed, the acceptance of a maximum likelihood estimate is, for me at least, qualified by the associated statistical problem, which is simplest in the standard large-sample case. Thus a statement on the support for any statistical hypothesis against another is a statistical inference; but how we make use of this result, for example, in a particular case, is not.

Edwards has proposed extending the range of support (so far defined in terms of likelihoods) to cover prior knowledge. He says 'The prior support for one hypothesis or set is S if, prior to any experiment, I support the one against the other to the same degree that I would if I had conducted an experiment which led to experimental support S in situation in which, prior to this conceptual experiment, I was unable to express any preference for the one over the other'. This proposal is an ingenious one which allows forms of inference intermediate between ones consistent with a more complete statistical attitude and with the Bayesian approach. However, while this new concept of prior support has by definition the numerical properties of experimental support, this

* A green ball is drawn from an urn which has three possible constitutions: (a) 99 green balls, 1 red; (b) 2 green, 98 red, (c) 1 green, 99 red. What is the best supported hypothesis: (a), (b) or (c)? (You are told that the frequency with which any colour is drawn, with replacement, from an urn is proportional to the number of balls of that colour in the urn.)

seems to me still reminisient of the Bayesian's introduction of prior probabilities with the numerical properties of statistical probabilities; in both cases, we cannot avoid the subjectivity introduced, as Edwards himself points out in the case of prior support. He gives an application of his ideas to a genetical example, where an inference about an unknown probability p of a mouse being homozygous for a certain gene locus depends on missing information about the type of parental mating. If two alternatives about this are allocated equal support, it proves possible to make two alternative statistical inferences A and B with a known numerical support in favour of one of them. While this inference is weaker (and more complicated) than the orthodox Bayes solution for the entire problem, it of course still hinges on the subjective allocation of equal prior support; indeed, Edwards' proposal is not limited to the use of *equal* prior supports, and the allocation of unequal prior supports would seem to me to have nearly as much arbitrariness as unequal prior probabilities.

My own predilection for statistical interpretation to inferences made in statistical situations is of course close to the behaviouristic outlook associated with Neyman. This standpoint has been the occasion of much ridicule from Bayesians and even Fisherians, but I will emphasize again (speaking for myself at least) that the recording of, say, a confidence statement does not imply that we confuse the inference for a unique occasion with a statistical inference, anymore than we confuse the statistical probability of an event with the realization at the next trial.

I do not propose to discuss here, in detail, the distinction between confidence and fiducial intervals, except to reiterate my personal view (1965), that the original fiducial interval concept introduced by Fisher was identical in interpretation with a class of confidence interval, and that it was only later that the statistical interpretation of a fiducial interval was specifically rejected by Fisher.

Returning to the recent emphasis on the use of the likelihood function as such, I would agree that this must (or course only if the statistical specification of the model is correct!) contain the relevant information on unknown parameters θ_i ($i = 1 \ldots k$), but I would also wish to know the sampling situation. For example, when it is proposed to 'accept' values of the parameters such that the likelihood ratio of the function at a parametric point θ_i $(i = 1 \ldots k)$ to the maximum value is not less than a fixed constant, how do we assign the constant? With the knowledge that $-2(L - L_{max})$ is in appropriate problems asymptotically a chi-square with degrees of freedom equal to K (L being log. likelihood), we can fix the critical value to correspond to a confidence region of known probability. Thsi is rather crude, and we can often improve on this considerably by a study of the derivatives $\partial L/\partial \theta_i$, but if this asymptotic theory is invalid it seems to me that the situation requires further specific appraisal.

Consider the following problem. It is admittedly of the kind I have labelled 'over-academic' but it does illustrate in an idealized form a natural epidemiological problem discussed by Almond (1954). A simple multiplicative birth-and-death process starting with n_0 individuals becomes extinct after N transitions. What can be inferred about the unknown parameter $p = \mu/(\gamma + \mu)$, where μ is the death-rate, γ the birth-rate? The likelihood function is fairly readily

$$Cp^{\frac{1}{2}(N + n_0)}q^{\frac{1}{2}(N - n_0)}, \tag{6.1}$$

which appears very well-behaved, but we must be very wary of the sampling situation. If $p > \frac{1}{2}$, all is well, and while N is random and not at our choice, the asymptotic sampling situation may be investigated in detail for n_0 not too small. However, if $p \leqslant \frac{1}{2}$, we have the remarkable result first established by O'N. Waugh (1958) that *conditional on extinction occurring* the distribution of N is identical with the case $p > \frac{1}{2}$ *with p and q interchanged.* The chance of extinction is $(p/q)^{n_0}$, and the likelihood function similarly factorizes to

$$(p/q)^{n_0}[Cq^{\frac{1}{2}(N + n_0)}p^{\frac{1}{2}(N - n_0)}]. \tag{6.2}$$

Thus any sampling theory would now yield a comparable inference about *q/p, given extinction*, coupled with the initial factor for the chance of extinction. This factor would of course be small for n_0 not too small, but it seems instructive to keep it separate, especially if the possibility of *selection* for a case of extinction could not be ruled out. For small n_0, the asymptotic theory would not be applicable, but confidence statements would still be available, even if rather complex in view of such results as (to take a trivial case with $n_0 = 1$ and $N = 1$

$$P(N > 1) < 0.1 \text{ if } p > 0.9;$$

and

$$P(1 < N < \infty | N < \infty) < 0.1,$$

with

$$P(N < \infty) < 0.11, \text{ if } p < 0.1.$$

Incidentally, the distribution of $P(N)$ for $n_0 = 1$ and $p > \frac{1}{2}$ i determined by its probability-generating function

$$\Pi(z) = [1-(1-4pqz^2)^{\frac{1}{2}}]/(2pq) \quad (6.3$$

with mean $2q/(p-q)$ and variance $4pq/(p-q)^3$.

Statistical properties of estimates

The view that it is important to know the statistical proportion of ou estimates may be emphasized in relation to another type of situatio examined in recent years, for example, in correction with th estimation of a multivariate mean by Stein (1962), but, as I pointed ou in the discussion on Stein's paper, effectively ventilated some time ag in the context of factor analysis.

In vector and matrix notation, let $z = M_0 f_0 + M_1 f_1$, where z is a se of test scores, f_0 the group factors and f_1 the specifics. We estimate f_0 for a particular person with scores z by the equation (1937)

$$\check{f}_0 = J^{-1} M_0' M_1^{-2} z,$$

where

$$\begin{cases} J = M_0' M_1^{-2} M_0, & E_1[\check{f}_0] = f_0, \\ E_1[(\check{f}_0 - f_0)(\check{f}_0 - f_0)'] = J^{-1}, & \end{cases}$$

E_1 denoting averaging over all possible sets of tests comparable with th given set in regard to the information on the group factors f_0.

By contrast, Godfrey Thomson's regression estimates (1939, p. 305 were specified by

$$f_0 = K\check{f}_0, K^{-1} = I + J^{-1},$$

where

$$\begin{cases} E[\hat{f}_0] = E[f_0] = 0, \\ E[(\hat{f}_0 - f_0)(\hat{f}_0 - f_0)'] = I - K, \end{cases}$$

E denoting averaging over all persons in the (standardized) population.

In the particular case of only one common factor, g, say, J reduces to the quantity

$$S = \Sigma_i r_{ig}^2/(1 - r_{ig}^2)$$

and we have

$$\sigma^2(\check{g}) = 1/S, \text{v}(\hat{g}) = 1/(1 + \text{S}) < \sigma^2(\check{g}),$$

where $\nu(\hat{g})$ refers to the 'mean square error' of $\hat{g}$, the appropriate measure for comparison because the $\hat{g}$ estimate is biased towards zero. Note that while $\nu(\hat{g})$ is less than $\sigma^2(\check{g})$, it is an average over persons of the quantity

$$\frac{g^2}{(S + 1)^2} + \frac{S}{(S + 1)^2},$$

and this will be greater than $\sigma^2(\check{g})$ for some persons. When the different statistical properties of $\hat{f}$ and $\check{f}$ are elucidated, it is possible to choose one in preference to the other in a given situation and for a given purpose. A concentration on one or more particular persons would favour $\check{f}$; a general investigation on the entire population would favour $\hat{f}$. However, as I pointed out at the time, one set is obtainable from the other, so that from the standpoint of the reduction of data they are still equivalent.

Conclusion

If I could sum up very briefly my own outlook, it is that statistics as a subject cannot from its very nature be concerned with single individuals or events as such, and consequently I do not see that *statistical* inferences can either. This is not to say that in the statistical domain the most relevant choice of statistical framework for the problem has not to be sought, nor that the inevitable transition from the statistical situation to a specific individual or event is to be overlooked. The first of these two tasks is still in the statistical domain. The second is not; statisticians are mere human beings when struggling with it, and should not confuse others by claiming otherwise.

References

Almond, J. (1954) A note on the χ^2 test applied to epidemic chains, *Biometrics*, **10**, 459.

Bartlett, H. S. (1933) Methods of estimating mental factors, *Nature*, **141**, 609–10.

——— (1965) R. A. Fisher and the last fifty years of statistical methodolgy. *J. Amer. Statist. Ass.*, **60**, 395–409.

——— (1966) Review of *Logic of Statistical Inference*, by I. Hacking., *Biometrika*, **53**, 631–3.

——— (1969a) Review of *Mathematical Ideas in Biology*, by J. Maynard Smith. *Nature*, **221**, 291.

——— (1969b) Probability and prejudice: correspondence, *Nature*, **221**, 786.

Edwards, A. W. F. (1969) Statistical methods in scientific inference. *Nature*, **222**, 1 33–7.

Hacking, I. (1965) *Logic of Statistical Inference*, Cambridge University Press.

Lindley, D. V. (1969) Probability and prejudice: correspondence, *Nature*, **221**, 594.

Maynard Smith, J. (1968) *Mathematical Ideas in Biology*, Cambridge University Press.

Skellam, J. G. (1969) Models, inference and strategy, *Biometrica*, **25**, 457.

Stein, C. M. (1962) Confidence sets for the mean of multivariate normal distribution. *J. Roy. Statist. Soc.* B, **24**, 265.

Thomson, Godfrey H. (1939) *The Factorial Analysis of Human Ability*, University of London Press.

Waugh, W. A. O'N. (1955) Conditional Markov processes, *Biometrika*, **45**, 241.

Epidemics

Reprinted from *Statistics; A Guide to the Unknown*,
edited by Judith M Tanur Mosteller et al.,
by permission of Holden-Day Inc., San Francisco

A classic book by the late Professor Greenwood, a medical statistician who was an authority on epidemics, has the title *Epidemics and Crowd Diseases*, which emphasizes the relevance of the population, or community, in determining how epidemics arise and recur. One of the first to realize the need for more than purely empirical studies of epidemic phenomena was Ronald Ross, better known for his discovery of the role of the mosquito in transmitting malaria to human beings. Since those early years, the mechanism and behavior of epidemics arising from infection that spreads from individual to individual, either directly or by an intermediate carrier, have been extensively studied,* but it is perhaps fair to say that only in recent years have some quantitative features become understood and even now much remains to be unravelled.

Mathematical models and data

The statistical study of epidemics, then, has two aspects – on one hand, the medical statistics on some infectious disease of interest and, on the other hand, an appraisal of the theoretical consequences of the mathematical model believed to be representative, possibly in a very simplified and idealized fashion, of the actual epidemic situation. If the consequences of the model seem to agree broadly with the observed characteristics, there is some justification for thinking that the model is on the right lines, especially if it predicts some features that had been

* Other pioneering workers in this field include W. Hamer, A. G. McKendrick, H. E. Soper, and E. B. Wilson.

unknown, or at least had not been used, when the model wa
formulated.

How do we build a mathematical model of a population unde
attack by an infectious agent? There is no golden rule for success. Som
feel that everything that is known about the true epidemic situatio
should be set down and incorporated into the model. Unfortunatel
this procedure is liable to provide a very indigestible hotchpotch wit
which no mathematician can cope, and while in these days of large-scal
computers there is much more scope for studying the properties c
these possibly realistic, but certainly complicated, models, there is sti
much to be said for keeping our model as simple as is feasible withou
too gross a departure from the real state of affairs. Let us begin then a
the other extreme and put in our ingredients one at a time.

We start with a community of individuals susceptible to th
infection, let us say a number S of them. We must also have som
infection, and we will confine our attention to the situation in whic
this can be represented by a number of individuals already infected an
liable to pass on the disease. The astute reader will notice that even
we are concentrating on epidemic situations with person-to-perso
infection, we perhaps ought not to amalgamate *infected* and *infectiv*
persons. Some people may be already infected, but not yet infective
some might not be infected, at least visibly, and yet be infective – s
called *carriers*. As we are considering the simplest case, however, w
merely suppose there is a number I, say, of infective persons. When, a
is to be hoped in real life, these persons recover, they may becom
resusceptible sooner or later (as for the common cold) or permanentl
immune, as is observed with a very high proportion of people in th
case of measles.

While these ingredients for our epidemic recipe are very basic an
common to many situations, a better check on our model is obtained
we are more definite and refer to just one disease; so let us conside
only measles from now on. This is largely a children's complain
mainly because most adults in contact with the virus responsible hav
already become immune and so do not concern us. Measles is no longe
as serious an illness as it used to be (even less now that there is
preventive vaccine), but is a convenient one to discuss because many c
its characteristics are fairly definite: the performance of subseque

immunity, the incubation period of about a fortnight, and the requirement of notifiability in several countries, including the U.S., England, and Wales. The last requirement ensures the existence of official statistics, though it is known that notifications, unfortunately, are far from complete. Provided we bear this last point in mind, however, and where necessary make allowance for incomplete notification, it should not mislead us.

A deterministic mathematical model

To return to our mock epidemic, we next suppose that the infective persons begin to infect the susceptibles. If the infectives remain infective, all the susceptibles come down with the infection eventually, and that is more or less all there is to be said. A more interesting situation arises when the infective persons may recover (or die, or be removed from contact with susceptibles) because then a competition between the number of new infections of susceptibles and the number of recoveries of infectives is set up. At the beginning of the epidemic, when there may be a large number of susceptibles to be infected, a kind of chain reaction can occur, and the number of notifications of new infected persons may begin to rise rapidly; later on, when there are fewer susceptibles, the rate of new notifications will begin to go down, and the epidemic will subside.

If we drew a graph of the number of infectives I against the number of susceptibles S at each moment, it would look broadly like the curve in Fig. 1. The precise path, of course, will depend on the exact assumptions made on the overall rate of infection, on whether this is strictly proportional to the number I, and on whether also proportional to S, so that the rate at any moment is, say, calculated from the formula aIS where a is a constant. The path will depend also on the rate of recovery of the infected population, likely to be proportional to I and given by bI, say, where b is a constant. Without going into too much detail yet, we can note one or two distinct features in the figure: (1) the susceptible population may not be reduced to zero at the time that the sources of infection have been eliminated; (2) because the path has no 'memory,' we could start at any point on it and proceed along the same curve – if our starting point were to the right of the

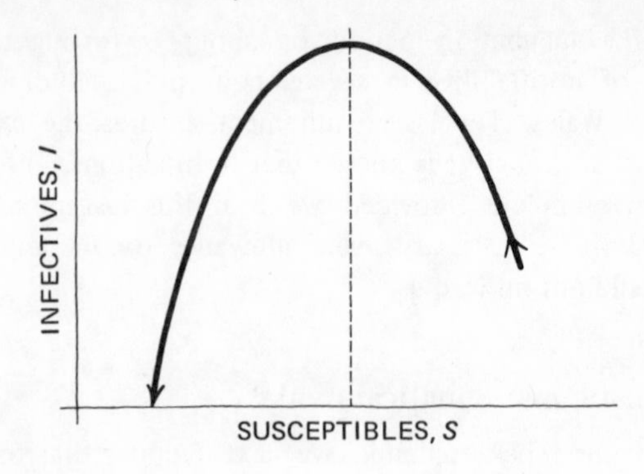

Fig. 1
General path of an epidemic beginning with many susceptibles S, increasing at first the number of infectives I, then decreasing

maximum point, our infectives I would rise, if to the left, they would fall.

Actually, our more detailed assumptions are equivalent to the pair of recurrence relations for calculating I_{t+1}, S_{t+1} at any time $t + 1$ in terms of the values I_t, S_t at the previous time t (the unit of time should be small, say, a day or week):

$$S_{t+1} = S_t - aI_tS_t, \qquad I_{t+1} = I_t + aI_tS_t - bI_t.$$

We notice that $I_{t+1} - I_t$ is positive or negative according to whether $aS_t - b$ is positive or negative. The value b/a for S (on our assumptions it is independent of I) is called the *critical threshold* for reasons that will become clear.

Thus far we have a model for a 'closed epidemic,' which terminates when I is zero. Can we turn it into a model for measles, which has been claimed to come in epidemics about every two years? One ingredient still missing is an influx of susceptibles, due, in the case of childhood illnesses, to births within the community.

Let us, therefore, add a term c, say, to the right-hand side of the first equation above. If we follow the course of events in Fig. 1, the path will start turning right before it reaches the axis $I = 0$ and can be shown o proceed in an ever-decreasing spiral (Fig. 2) till it finally arrives at an quilibrium point, which is determined by the equations

$$c - aIS = 0, \qquad aIS - bI = 0.$$

The second of these gives $S = b/a$ (the critical threshold value), and the irst then yields $I = c/b$. These results are partly encouraging, but partly rroneous. The encouraging feature is the tendency to recurrent pidemics; we can even find the period of the cycle, which is found to e approximately

$$\frac{2\pi}{\sqrt{(ac - \frac{1}{4}a^2c^2/b)}}$$

Sir William Hamer, who first put forward the above model for measles n 1906, took $b = \frac{1}{2}$ when t is reckoned in weeks, corresponding to an verage incubation period of a fortnight, and c for London at that time s 2200. The value of a is more uncertain, but one method of arriving at t is to note that the *average* number of susceptibles, which was put at 150 000, should be around the theoretical equilibrium value b/a giving $a = 1/300\,000$. Notice that c will tend to be proportional to the size of he community, so that if ac is to remain constant, a must be inversely

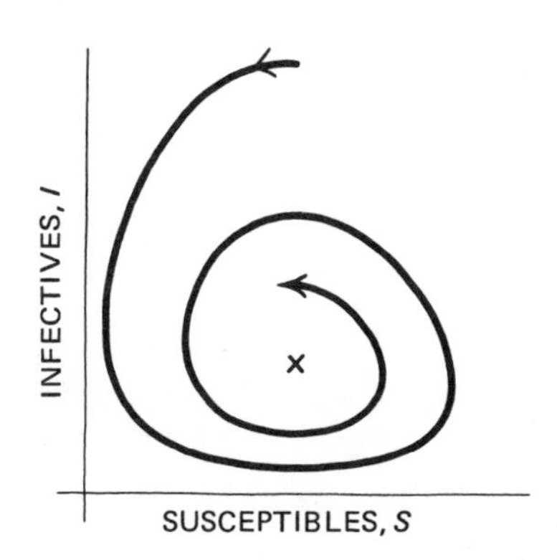

Fig. 2
Deterministic model. Approach to equilibrium point (at cross) of I,S curve

proportional to the population, but this is not an unreasonable assumption; it is consistent, for example, with effective infectivity over a constant urban area, the entire town being regarded as an assemblage of such units.

The introduction of an influx of susceptibles showed that instead of following the simply curved path of Fig. 1, an epidemic might follow a spiral until it finally settled down with a particular number of susceptibles. The time to go around the spiral once, called the *period*, is estimated for London data at 74 weeks, in reasonable agreement with the average period of somewhat less than two years that has been observed for large towns in England and Wales (see Table 2), the U.S. and comparable countries in the present century. We would hardly expect the epidemic pattern to remain precisely the same under very different social conditions, though the annual measles mortality figures for London quoted from John Graunt for the seventeenth century (Table 1) suggest a similar pattern even then (with perhaps a slightly longer average period of 2 to 3 years).

Table 1. Deaths from measles in London in the seventeenth century

1629	*1630*	*1631*	*1632*	*1633*	*1634*	*1635*	*1636*	*1637–46*	*1647*	*1648*
41	2	3	80	21	33	27	12	Not recorded	5	92

1649	*1650*	*1651*	*1652*	*1653*	*1654*	*1655*	*1656*	*1657*	*1658*	*1659*	*1660*
3	33	33	62	8	52	11	153	15	80	6	74

Unsatisfactory features of the model

The erroneous feature of the improved model is that actual measles in London or other large towns recurs in epidemics without settling down to a steady endemic state represented by the theoretical equilibrium point. What aspects of our model must we correct? There are some obvious points to look at:

(1) Our assumptions about the rate of recovery correspond to a gradual and steadily decreasing fraction of any group infected at the same time whereas the incubation period is fairly precise at about two weeks.

before the rash appears and the sick child is likely to be isolated (this being equivalent to recovery).

(2) We have ignored the way the children are distributed over the town, coming to school if they are old enough or staying at home during a vacation period.

(3) Measles is partly seasonal in its appearance, with a swing in average notifications from about 60% below average in summer to 60% above average in winter.

Introducing chance into the model

We will consider these points in turn. The effect of point (1) is to lessen the 'damping down' to the equilibrium level, but not, when correctly formulated, to eliminate it. Point (2), on the movement over the town, raises interesting questions about the rate of spread of infection across different districts, but is less relevant to the epidemic pattern in time, except for its possible effect on (3). If we postulate a ±10% variation in the 'coefficient of infectivity' a over the year, it is found to account for the observed ±60% or so in notifications. There seems to be little evidence of an intrinsic change in a due, say, to weather conditions, and it may well be an artifact arising from dispersal for the long summer vacation and crowding together of children after the holidays. Whatever its cause, it does not explain the persistence of a natural period; only the seasonal variation would remain and give a strict annual period, still at variance with observation.

To proceed further, let us retrace our steps to our closed epidemic model of Fig. 1. To fix our ideas, suppose we initially had only one infective individual in the community. Then the course of events is not certain to be as depicted; it may happen that this individual recovers (or is isolated) before passing on the infection, even if the size of susceptible population is above the critical threshold. This emphasizes the chance element in epidemics, especially at the beginning of the outbreak, and this element is specifically introduced by means of probability theory. To examine the difference it makes, let us suppose the *chance* P of a new infection is now proportional to aIS and the chance Q of a recovery proportional to bI. Denote the chance of the outbreak ultimately fading out without causing a major epidemic by p.

We shall suppose also that the initial number S_0 of susceptibles is large enough for us not to worry about the proportionate change in S if the (small) number of infective persons changes. Under these conditions two infective persons can be thought of as acting independently in spreading infection, so that the chance of the outbreak fading out with *two* initial infective persons must be p^2.

Now consider the situation after the first 'happening.' Either this is a new infection or a recovery, and the relative odds are $P/Q = aS_0/b$. If it is a new infection, I changes from 1 to 2, and the chance of fade-out is p^2 from now on, or I drops to zero, and fade-out has already occurred. This gives the relation

$$p = \frac{P}{P+Q}p^2 + \frac{Q}{P+Q},$$

whence either $p = 1$ or $p = Q/P = b/(aS_0)$. If $b \geqslant aS_0$ (that is, if we are below the critical threshold) the only possible solution is $p = 1$, implying as expected that the outbreak certainly fades out. However, the ultimate probability of fade-out can be envisaged as the final value reached by the probability of fade-out up to some definite time t, this more general probability steadily increasing from zero at $t = 0$ to its limiting value, which therefore will be the *smaller* of the roots of the above quadratic equation. This is $b/(aS_0)$ if this value is less than one, providing us with a quantitative (nonzero) value of the chance of fade-out *even if the critical threshold is exceeded* and stressing the new and rather remarkable complications that arise when probabilistic concepts are brought in.

The mathematics shows that if the initial number of susceptibles is smaller than a value determined by the ratio of some rates used in the model, then the epidemic will certainly fade out. If it is larger than this critical value, then there is still a positive probability that the epidemic will fade out.

When new susceptibles are continually introduced, represented by c, the complications are even greater. For small communities, however, the qualitative features can be guessed. Once below the threshold, the number of infectives will tend to drop to zero, and though the susceptibles S can increase because of c, it seems unlikely that the number will pass the threshold before I has dropped to zero. The epidemic is now finished, and cannot re-start unless we introduce some *new* infection from outside the community. This is exactly what is observed with measles in a small isolated community, whether it is a boarding school, a rural village, or an island community. For such communities, the period between epidemics depends partly on the rate of immigration of new infection into the area and not just on the natural epidemic cycle. Moreover, when new infection enters, it cannot take proper hold if the susceptible population is still below the threshold, and even if above, new infection may have to enter a few times before a major outbreak occurs. The average period tends to be above the natural period for such communities. If we assume that the rate of immigration of new infection is likely to be proportional to the population of the community, the average period between epidemics will tend to be larger the smaller the community, and this again is what is observed (Table 2).

Consider now a larger community. We expect random effects to be proportionately less; there is still, however, the possibility of extinction when the critical threshold is not exceeded. Nevertheless, before all the infectives have disappeared, the influx c of new susceptibles may have

Table 2. Measles epidemics for towns in England and Wales (1940–1956)

Town	*Population (thousands)*	*Mean period between epidemics (weeks)*	*Town*	*Population (thousands)*	*Mean period between epidemics (weeks)*
Birmingham	1046	73	Newbury	18	92
Manchester	658	106	Carmarthen	12	79
Bristol	415	92	Penrith	11	98
Hull	269	93	Ffestiniog	7.1	199
Plymouth	180	94	Brecon	5.6	149
Norwich	113	80	Okehampton	4.0	105
Barrow-in-Furness	66	74	Cardigan	3.5	>284
Carlisle	65	75	South Molton	3.1	191
Bridgewater	22	86	Llanrwst	2.6	>284
			Appleby	1.7	175

Source: Bartlett (1957), Tables 1 and 2.

swung S above the threshold, and the stage is set for a new epidemic. Under these conditions (and provided ac remains constant, as already assumed), the natural period will change little with the size of community.

Critical size of community

How large does the community have to be if it is to begin to be independent of outside infection and if its epidemic cycle is to be semipermanent? Exact mathematical results are difficult to obtain, but approximate solutions have been supplemented by simulation studies of the epidemic model, using computers. An example of one such series plotted to extinction of infection after four epidemics (an interval representing nearly seven years) is shown in Fig. 3. This particular series has no built-in seasonal incidence, but some internal migration within its population boundaries; its average susceptible population is 3700. It appears from all such results that the critical size of the susceptible population is, for the measles model, of the order of 7000, or if the factor of 40 estimated for Manchester, England, between total and susceptible population is used, over a quarter of a million people in the community.

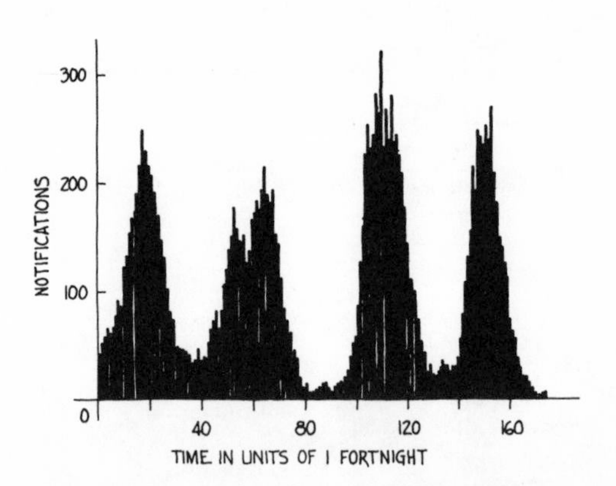

Fig. 3

Results of simulation of four epidemics of measles over a seven-year period for a town whose average susceptible population is 3700

Now we do not need to use this theoretical figure as more than an indication of what to look for. By direct examination of measles notifications for any town, we can see whether notifications have been absent for more than two or three weeks. In view of the rather well-defined incubation period, we would infer from this lack of notifications that the infection had disappeared if we knew that notifications were complete. Incomplete notification is a complication, but not one that is likely to affect these quantitative conclusions very much, for fade-out of infection is found to increase rather rapidly as the community size decreases and soon becomes quite recognizable from the detailed statistics. In this way, it was ascertained that in England and Wales, during the period 1940–1956, cities of critical size were Bristol (population about 415 000) and Hull (269 000). This investigation was supplemented by an examination of U.S. statistics for the period 1921–1940, from which it was found that some comparable North American cities were Akron (245 000), Providence (254 000), and Rochester (325 000). Therefore, there is an observed critical community size of about 300 000, in reasonable agreement with what we were expecting.

Of course, towns of such size are not completly isolated from other communities as assumed in our model; this could tend to lessen the observed critical size, especially if the isolation is comparatively slight. In Table 3 the fade-out effect is shown for aggregates of individual

Table 3. Observed (aggregate) Fade-out effect in Manchester wards

Wards	*Cumulative population (thousands)*	*Number of epidemics followed by fade-out*	*Probability of fade-out (%)*
Ardwick	18.4	12	100
St. Mark's	38.2	12	100
St. Luke's and New Cross	71.8	9	75
All Saints, Beswick, and Miles Platting	140.8	4	33
Openshaw, Longsight, N. and S. Gorton, Bradford, and St. Michaels	254.1	1	8
Medlock St., W. and E. Moss Side, Rusholme, Newton Heath, Collyhurst, Harpurhey, and Cheetham	419.3	0	0

Source: Bartlett (1957), Table 3.

'wards' in Manchester to demonstrate how it decreases with the population aggregate. The critical size (defined precisely in terms of 50% probability of fade-out after an epidemic) is, again as expected smaller than for complete towns due to the extensive migration across the ward boundaries; it is estimated to be 120 000 total population living within the area.

Conclusions

If we review these results, we may justifiably claim that our theoretical model for measles, idealized though it inevitably is, has achieved some fair degree of agreement with the observed epidemic pattern. In particular:

(1) It predicts a 'natural' period between epidemics of rather less than two years.
(2) A small (±10%) seasonal variation in infectivity (whether or not an artifact of seasonal pattern in school-children's movements) accounts for the larger (±60%) observed seasonal variation in notifications.
(3) It predicts extinction of infection for small communities, with consequent extension (and greater variability) of periods between epidemics.
(4) It predicts a critical community size of over a quarter of a million necessary for the infection to remain in the community from one epidemic to the next.

Epidemic patterns, of course, will be very sensitive to changing customs and knowledge; and the introduction of a vaccine for measles will inevitably change its epidemic pattern, and perhaps in time eliminate the virus completely. However, the greater understanding of epidemics that follows from appropriate models may be applied to other epidemic infections, and should assist in predicting and assessing the consequences of any changed medical practice or social custom even for measles.

References

M. S. Bartlett. (1957) 'Measles Periodicity and Community Size.' *Journal of the Royal Statistical Society*, A120. 48–70.

EPIDEMICS

J. Graunt. (1662) Reprint, 1939. *Natural and Political Observations upon Bills of Mortality*. Johns Hopkins University Press, Baltimore.

M. Greenwood. (1935) *Epidemics and Crowd Diseases.* William and Norgate, London.

R. Ross. (1910) *The Prevention of Malaria*, Second Edition. J. Murray, London.

Equations and models of population change

Reprinted from the *Mathematical Theory of the Dynamics of Biological Populations*, published by Academic Press on behalf of the Institute of Mathematics and its Applications

1. Remarks on Theoretical Model-Building

The mathematician and the biologist have an almost inevitable confrontation in the formulation of a model for any biological situation, for, whereas the biologist will probably begin with the actual and complex set of facts that he has observed or otherwise become aware of, the mathematician is likely to be looking for an idealized model capable of theoretical investigation. This kind of dilemma has become more obvious as the use of mathematics has moved from the physical to the biological (and also to the social) sciences, and I do not believe that there is a facile answer. It is quite natural for an experimental or observational scientist to list a large collection of facts that should be taken into account in any comprehensive theoretical formulation; but the purpose of the latter is to understand and interpret the facts, and the mathematician is sometimes able to give insight by a study of idealized and over-simplified models, which are more complicated in their behaviour than is often realized. Until we can understand the properties of these simple models, how can we begin to understand the more complicated real situations? There is plenty of scope for variety of approach, but I would urge that critical criteria of success be employed wherever possible. These include the extent to which:

(i) known facts are accounted for;

(ii) greater insight and understanding are achieved of the biological situation being studied;

(iii) the theory or model can correctly predict the future pattern, even under different conditions from those pertaining to the current observed data.

Over-simplified models are obvious targets for criticism, but model-builders who are more ambitious in their attention to detail must bear in mind two often relevant points:

(a) If the simplified model has retained the basic features of the real situation, there may well be "diminishing returns" as further parameters and features are introduced;

(b) The more parameters, the more problems of estimation, and this can seriously add to the difficulties of building what may be in principle a more realistic model.

2. Equations of Population Change

Let me try next to survey some of the basic equations of population change, and see how they fit into the historical picture. (They are applicable to physical or chemical situations as well as biological but I shall have the biological in mind.) It is convenient and feasible to formulate these equations in a general probabilistic or *stochastic* framework, without implying by this that the so-called *deterministic* framework is not adequate in many contexts.

I shall denote a set of population sizes at time t of various interacting species or types (including if necessary different classes by age-groups or sex or location or genetic composition of the same species) by the mathematical vector notation $\mathbf{N}_t$ where t may be discrete (generation number) or continuous. For definiteness we shall mainly consider the latter case in this introductory presentation, though analogous formulations in discrete time have been familiar from the work of mathematical geneticists (in relation to successive *generations*), or in population theory from the work of P. H. Leslie (1945), leading to matrix recurrence equations which are in particular convenient for simulation calculations.

In most contexts it is possible to classify the process as what is called *Markovian*, by which is meant that the future evolution of the process can be calculated in principle from a knowledge of the population vector $\mathbf{N}_t$ at any given time t. This is obviously not valid in general if $\mathbf{N}_t$ omits specification by, say, age and sex, but may be valid if such classifications are included. An example which appears to be an exception is that of human populations increasing under a social custom of family planning, but even such examples may, at least in principle, be included by appropriate redefinition, such as counting married couples as units, classified by their family size. Even at this stage it is evident how the mathematical formulation can increase in complexity with the realism of the specification, but not necessarily with a change in the basic mathematical framework. (There is an additional technical complication in the exact handling of age increase or spatial movement, owing to the more complicated nature of $\mathbf{N}_t$ when it involves subdivision by ancillary *continuous* variables, but I will come back to that later.)

In continuous time, the probabilities of changes are reasonably assumed proportional to the time interval $(t, t + \Delta t)$ for small enough Δt (thereby ensuring a non-zero probability of change in a non-zero interval $t, t + \tau$, and a zero probability of change in a zero interval). Thus there is a transition probability function

$$R\{\mathbf{N}_{t+\Delta t}|\mathbf{N}_t\}\Delta t \text{ of order } \Delta t \text{ for } \mathbf{N}_{t+\Delta t} \neq \mathbf{N}_t,$$

by means of which the probabilistic differential equation (known as the Kolmogorov "forward" equation) governing the changes in $\mathbf{N}_t$ from t to $t + \Delta t$ may be written down.

It is often mathematically convenient, as I pointed out many years ago (e.g. Bartlett, 1949), to formulate these changes in one "omnibus" equation for what is called the probability-generating (or some equivalent) function of $\mathbf{N}_t$, as shown in more detail in the Appendix.

This equation (for, say, the probability-generating function $\Pi_t(\mathbf{z})$) may be written

$$\frac{\partial \Pi_t(\mathbf{z})}{\partial t} = H_t \Pi_t(\mathbf{z}),$$

where H_t is an *operator* (see Appendix I) acting on $\Pi_t(\mathbf{z})$. In general it will involve t explicitly, as in seasonal variation of birth- or death-rates, but if it is what we term *homogeneous* in time, then the above equation has the simple formal solution

$$\Pi_t(\mathbf{z}) = \mathrm{e}^{Ht}\Pi_0(\mathbf{z}),$$

though this equation must not mislead us into thinking that even some of the most elementary natural examples of it, especially those involving more than one type of individual, have other than complicated or intractable solutions.

In the time-homogeneous case there may be an equilibrium situation, which will then satisfy the equation

$$H\Pi(\mathbf{z}) = 0.$$

Sometimes it is illuminating to formulate this general model in direct stochastic terms. We have postulated a stochastic change in time Δt, $\Delta\mathbf{N}_t = \mathbf{N}_{t+\Delta t} - \mathbf{N}_t$, with probability $R\{\mathbf{N}_{t+\Delta t}|\mathbf{N}_t\}\Delta t$, and this specification is equivalent to some equation

$$\mathrm{d}\mathbf{N}_t = f(\mathbf{N}_t)\,\mathrm{d}t + \mathrm{d}\mathbf{Z}_t,$$

where $\mathrm{d}\mathbf{Z}_t$ has zero mean and known distributional properties. This equation has its *deterministic* analogue

$$\frac{\mathrm{d}\mathbf{n}_t}{\mathrm{d}t} = f(\mathbf{n}_t),$$

which neglects $\mathrm{d}\mathbf{Z}_t$.

When the solutions of these equations are too intractable, as is in any case likely when further practical details are inserted, it will often be more feasible and instructive to compute numerical *realizations* of the process. In the deterministic version this is in fact identical with the numerical solution (for given starting conditions); but note that in the stochastic version several simulations must be computed in order to have any idea of the complete probabilistic situation, and the known complete solutions in some relatively simple cases have warned us that these can sometimes be rather complex. It is important in more realistic formulations to judge whether stochastic effects are likely to be important, and this judgement in turn can be assisted by theoretical studies of some of the simpler cases.

The remainder of this paper refers to three examples in different biological fields: epidemiological, genetical and demographic. Firstly, some of the properties of the stochastic model for recurrent epidemics in a population with a steady influx of susceptible individuals, which I have previously employed (see, for example, Bartlett, 1960) as a model for measles epidemics, are recalled, and also the *predicted* changes in these properties following the current programme of *partial* vaccination against measles infection, recently investigated by D. A. Griffiths (1971).

3. The Recurrent Epidemics Model for Measles

A theoretical model for measles epidemics goes back to the (deterministic) formulation by W. Hamer in 1906, and has since been discussed extensively by various writers, so that full details and background should be sought in the previous literature. To recapitulate briefly, however, on the model, this represents a population of susceptible individuals in homogeneous contact with a number of infected and infective individuals, the latter being removed (or permanently recovering) analogously to a mortality, and a steady influx of susceptibles represented in the measles application by real births in the community. An influx of infective individuals may also be included if relevant. When the community is closed to such infection from outside, the only parameters are the removal or "mortality" rate μ for the infectives, depending on the incubation period, the influx rate ν for susceptibles, depending on the community population size and the birth-rate, and the infectivity coefficient λ, or equivalently the quantity $\rho = \lambda\nu/\mu$. The value of the latter was originally estimated by Hamer for London as $1/68{\cdot}2$, though Griffiths (*loc. cit.*) has suggested that a smaller value for ρ may be more appropriate (even prior to the vaccination era).

My reasons for referring to this model here (for a summary of its mathematical formulation and properties, see Appendix III) are partly because it is an example of a model which, in spite of its simplified character as formulated above (this neglects seasonal periodicity, the discrete aspects of the incubation period for infection, and spatial heterogeneity, all of which can of course be introduced, especially into simulation studies) has the three features referred to earlier which make it a useful and illuminating model. The earlier deterministic versions of the model gave the oscillatory tendency correctly, but greater insight came from a realization of the importance of "fade-out" of infection associated with the stochastic model, leading to quantitative studies of the concept of a critical population size (around a quarter of a million), and of the inverse relation of epidemic periodicity with community size for communities below this critical size.

This success with the model justifies moreover the claim that it should be relevant even when the recent introduction of vaccine changes the conditions drastically. D. A. Griffiths (*loc. cit*; see also Appendix III), on examining this problem has drawn some conclusions which seem rather disturbing, namely, that a *partial* vaccination programme, which reduces but does not eliminate the susceptible influx, will merely increase the intra-epidemic period without lessening the intensity of epidemics; and, moreover, such an increase in period will if anything put up the average age of onset for those who become infected, with a consequent increased risk to them of complications arising from the infection.

4. The Temporal History of a Gene Mutation

The second example I wish to refer to represents a rather small, but nevertheless very vital, part of the theoretical problems of population genetics: the fate of a gene mutation. Here my purpose is again to show that simple models, if correctly used, can be informative, perhaps more so than has always been appreciated. A useful account of this problem will be found in Chapter 1 of Kimura and Ohta (1971); my remarks are intended by way of further commentary.

In addition to the usual simplifications such as random mating (in the case of sexual reproduction for diploids) there is, as Kimura and Ohta note, a choice of mathematical technique. Either we may concentrate on the early stages of the mutant gene, when its stochastic history may be dealt with by the theory of branching processes (see Appendix I, example (a)) neglecting the finite size of the population; or we may consider the later stages, when the frequency ratio of the number of mutant genes at that locus in the population to the total number of genes at that locus, is more relevant, and we treat it as effectively a continuous variable changing according to a diffusion equation (Appendix I, example (b)).

In fact, it seems advisable to make use of both techniques, so that any limitations of one are high-lighted by the other, and our conclusions are unlikely to be so restricted by the theoretical techniques used. To be more specific, let us consider a neutral mutant gene in a finite population of fixed size N (total gene number $2N$ for a diploid). If the initial number of such mutant genes is not too small, the diffusion equation may be used to obtain not only the probability of extinction or fixation of the mutant (the latter probability being $n/(2N)$ when n is the initial number of mutant genes) but also the distribution of the time to extinction or fixation. Formulae for the mean and variance of this distribution have been given by Kimura and Ohta (cf. also Ewens, 1963); in particular, the mean time conditional on fixation is approximately $4N$ (independent of n), compared with $2n \log (2N/n)$ to extinction. (In the Appendix a rather more general formulation is given.)

However, let us now consider the branching process approach, even at the cost of making N infinite. We start with just one mutant ($n = 1$), but realize that the number of mutant genes in the progeny from this mutant is very relevant. Indeed, we get a different distribution with different assumptions, the probability of extinction at the first generation being e^{-1} if the number of mutant genes has a Poisson distribution with mean one, but $\frac{1}{2}$ if the number of progeny from a mating is restricted to two. The complete distribution of the time to extinction may be investigated (cf. Nei, 1971a), but in view of these difficulties let us simplify further by taking time continuous (as in the diffusion approximation), and represent the process by a simple birth-and-death process (with $\lambda = \mu$ for the neutral case). The value of the parameter λ determines the time scale and may be adjusted to give any desired probability of extinction at $t = 1$ (one generation), consistently with the arbitrariness already referred to.

My justifications for this somewhat naïve suggestion are:

(i) The resulting distribution agrees well with simulation results reported by Kimura and Ohta (1969b), a little better for $\lambda = e^{-1}$ than $\lambda = \frac{1}{2}$, but in neither case is χ^2 significant.

(ii) The distribution is known to have infinite moments, and, while this is in effect a consequence of N infinite, it is sufficient to warn us that an appeal to means and variances may be most misleading for some of the extremely J-shaped distributions arising in this context.

I would in fact suggest that Kimura's earlier (1964) solution for the cumulative distribution (see Appendix II) is more relevant, and that percentiles should be used in this context in preference to moments. Another problem where this suggestion is relevant is that of the total number of new mutant individuals before the mutant type finally dies out (as it does eventually if N is infinite); here again the theoretical mean is infinite in the neutral case. Even in the case of a deleterious mutant (for example, the interesting application by Nei, (1971b), to the haemoglobin mutant Hb M_{IWATE}) the distributions of extinction time and total mutant numbers may be extremely skew. Incidentally the continuous time birth-and-death adaptation to this problem must keep track of new mutant individuals as new "births", but this is technically feasible (Appendix II).

5. Auxiliary Variables such as Age

Coming finally to a few remarks on relevant equations in the general area of demography, I will first of all recall that equations of population growth were classically formulated, for example, by the American actuary A. J. Lotka (see Keyfitz, 1968) in deterministic terms, the growth and change of the population being developed theoretically or numerically from its detailed age composition and its mortality and birth-rates for individuals of different ages. (The complication of the two sexes is a further technical difficulty.) This formulation is sufficient for most actuarial and demographic purposes; but it is as well to realize the theoretical need for the more comprehensive stochastic formulation by noticing that any finite population can merely consist of a number of individuals of specific ages (and sex), and asking what the relation is of such a population with theoretical concepts such as stable age distributions and ultimate exponential growth.

The answer for many of the more straightforward models (in fact, "linear" models) of population growth is that these theoretical results refer to *expected* values, and fluctuations from these expected values may, if desired, also be investigated. The complete stochastic formulation is a matter of some sophistication (see D. G. Kendall, 1949), but may often be relevant if only to ensure that when we refer to, say, an age distribution we are speaking about a property of our model which makes theoretical sense. Sometimes, however, this wider approach enables us to ask relevant questions which cannot even be formulated in a deterministic framework. This has obviously been true in the epidemiological and genetic examples already discussed, but if we refer to the derivation (see Bartlett, 1970, or Appendix IV) of stable age distributions not only in the standard population case but also in the more recent application to proliferating cell populations (Cleaver, 1967) we may notice several queries concerning the stochastic independence of successive phase-times of a dividing cell, the stochastic independence of subsequent daughter cells from the same parent, and so on, some or all of which may affect the analysis and interpretation of particular experimental results (cf. Macdonald, 1970).

Mathematical Appendix

I. Basic Equations

Population number N_t at time t; or, more generally, vector $\mathbf{N}_t$, or even $\mathbf{N}_t(u)$, where u may be age or position.

Changes in N_t from 0 to t assumed governed by matrix

$$\prod_{i=1}^{t} \mathbf{Q}_i$$

where $\mathbf{Q}_i$ matrix of transition probabilities between times i and $i+1$. In time-homogeneous case, this is $\mathbf{Q}^t$, or equivalently $e^{\mathbf{R}t}$ in continuous time case.

If we define (in latter case)

$$\pi_t(\mathbf{z}) = M_t(\boldsymbol{\theta}) = E\{\exp \boldsymbol{\theta}'\mathbf{N}_t\}, \qquad (\boldsymbol{\theta} = \log \mathbf{z}),$$

then (Bartlett, 1949)

$$\frac{\partial M_t(\boldsymbol{\theta})}{\partial t} = H\left(\boldsymbol{\theta}, t, \frac{\partial}{\partial \boldsymbol{\theta}}\right) M_t(\boldsymbol{\theta}), \tag{1}$$

where

$$H(\boldsymbol{\theta}, t, \mathbf{n}) = \lim_{\Delta t \to 0} E\left\{\frac{\exp(\boldsymbol{\theta}'\Delta\mathbf{N}_t) - 1}{\Delta t}\middle| \mathbf{N}_t = \mathbf{n}\right\},$$

or equivalently in terms of $\pi_t(\mathbf{z})$.

In time-homogeneous case, equilibrium solution (if it exists) satisfies

$$H\left(\boldsymbol{\theta}, \frac{\partial}{\partial \boldsymbol{\theta}}\right) M(\boldsymbol{\theta}) = 0. \tag{2}$$

Direct stochastic representation, say

$$\mathrm{d}\mathbf{N}_t = \mathbf{f}(\mathbf{N}_t)\,\mathrm{d}t + \mathrm{d}\mathbf{Z}_t,$$

where $E\{\mathrm{d}\mathbf{Z}_t\} = 0$, with deterministic approximation

$$\frac{\mathrm{d}\mathbf{N}_t}{\mathrm{d}t} = \mathbf{f}(\mathbf{N}_t). \tag{3}$$

Equilibrium points in (3) determined by

$$\mathbf{f}(\mathbf{N}_t) = 0. \tag{4}$$

Examples

(a) Multiplicative or branching processes
Discrete time:

$$\mathbf{z} \to \mathbf{G}(\mathbf{z})$$

$$\pi_{t+1}(\mathbf{z}) = \pi_t(\mathbf{G}(\mathbf{z})),$$

or equivalently ("backward" equation, with $\boldsymbol{\pi}_0(\mathbf{z}) = \mathbf{z}$)

$$\boldsymbol{\pi}_{t+1}(\mathbf{z}) = \mathbf{G}(\boldsymbol{\pi}_t(\mathbf{z})). \tag{5}$$

In continuous time, let

$$\mathbf{G}(\mathbf{z}) \to \mathbf{z} + \mathbf{g}(\mathbf{z})\,\mathrm{d}t,$$

then (5) becomes

$$\frac{\partial \boldsymbol{\pi}_t(\mathbf{z})}{\partial t} = \mathbf{g}(\boldsymbol{\pi}_t(\mathbf{z})). \tag{6}$$

(b) Normal diffusion approximation (one variable X_t)

$$H(\theta, t, X_t) \sim \mu(t, X_t)\theta + \tfrac{1}{2}v(t, X_t)\theta^2.$$

For probability density $f_t(x)$,

$$\frac{\partial f_t(x)}{\partial t} + \frac{\partial}{\partial x}[\mu(t, x) f_t(x)] = \tfrac{1}{2}\frac{\partial^2}{\partial x^2}[v(t, x) f_t(x)]. \tag{7}$$

From (7), in equilibrium (time-homogeneous case),

$$f(x) = Ah(x) \qquad \text{(Sewall Wright formula)} \tag{8}$$

where

$$h(x) = 1/[v(x)Q(x)],\ Q(x) = \exp\left\{-2\int_0^x \mu(u)\,du/v(u)\right\}$$

II. First Passage (e.g. extinction) Probabilities and Times

For absorption (fixation) problems in time-homogeneous case, consider "backward" equation equivalent to (7) viz if $f_t(x|x_0)$ is written $f_t(x_0)$ for brevity,

$$\frac{\partial f_t(x_0)}{\partial t} - \mu(x_0)\frac{\partial f_t(x_0)}{\partial x_0} - \tfrac{1}{2}v(x_0)\frac{\partial^2 f_t(x_0)}{\partial x_0^2} = 0 \tag{9}$$

Write

$$F_t(x) = \int_{-\infty}^{x} f_t(x), \qquad P_0(t) = F_t(0), \qquad P_1(t) = 1 - F_t(1),$$

$$P(t) = P_0(t) + P_1(t),$$

$$L(\psi) = L_0(\psi) + L_1(\psi) = \int_0^{\infty} \mathrm{e}^{-\psi u}\left[\frac{\partial P(u)}{\partial u}\right] \mathrm{d}u,$$

then distribution of first passage time to 0 or 1 determined by (cf. Cox and Miller, 1965, § 5.10)

$$\tfrac{1}{2}v(x_0)\frac{\partial^2 L_i}{\partial x_0^2} + \mu(x_0)\frac{\partial L_i}{\partial x_0} = \psi L_i, \qquad (L_i(x_0 = j) = \delta_i^j). \tag{10}$$

Solution of (10) well-known (by other methods) if v & μ constant. In more general case, one method of solution (when moments exist) is to write

$$L_i = \sum_{r=0}^{\infty} (-\psi)^r L_i^{(r)}(x_0)/r!,$$

then

$$L_i^{(0)} = P_i, \qquad P_1 = \int_0^{x_0} Q(u)\,du \bigg/ \int_0^1 Q(u)\,du, \qquad P_0 = 1 - P_1,$$

and

$$L_i^{(r)} = \left[\int_0^1 Q(u)\,du\right]\left[P_1 \int_{x_0}^1 2rP_0hL_i^{(r-1)}\,du + P_0 \int_0^{x_0} 2rP_1hL_i^{(r-1)}\,du\right] \tag{11}$$

In particular,

$$L_i^{(1)} = \left[\int_0^1 Q(u)\,du\right]\left[P_1 \int_{x_0}^1 2P_iP_0h\,du + P_0 \int_0^{x_0} 2P_iP_1h\,du\right] \tag{12}$$

(cf. Ewens, 1963; Kimura and Ohta, 1969a, b).

Alternatively, we may solve for $P_i(t)$ directly. For example, in case $\mu(x) = 0$, $v(x) = \kappa x(1 - x)$,

$$P_0(t) = 1 - x_0 + \frac{1}{2}\sum_{i=1}^{\infty} (-1)^i[L_{i-1}(2x_0 - 1) - L_{i+1}(2x_0 - 1)]\,e^{-\frac{1}{2}\kappa i(i+1)t} \tag{13}$$

where $L_i(x)$ are Legendre polynomials (Kimura, 1964), $P_1(t|x_0) = P_0(t|1 - x_0)$. Asymptotically,

$$P_0(t) \sim 1 - x_0 - 3x_0(1 - x_0)\,e^{-\kappa t}. \tag{14}$$

From (5) or (6), for $G(z)$ or $g(z)$ (one type of individual), extinction probability $\pi_t(0)$. (In discrete time, if $G(z) = \exp(z - 1)$, $\pi_1(0) = e^{-1}$; if $G(z) = \frac{1}{2}(1 - z^2)$, $\pi_1(0) = \frac{1}{2}$).

For simple birth-and-death process in continuous time,

$$g(z) = (\lambda z - \mu)(z - 1)$$

and

$$\pi_t(0) = \frac{\mu(T-1)}{\lambda T - \mu}, \qquad T = e^{(\lambda-\mu)t}. \tag{15}$$

($\lambda < \mu$; or, $\lambda > \mu$, *given extinction*, interchange λ and μ). For this process, distribution of time to extinction given by

$$\left[\frac{\partial \pi_t(0)}{\partial t}\right] \mathrm{d}t = \frac{\mu(\mu-\lambda)\,\mathrm{d}T}{(\lambda T - \mu)^2}, \qquad (T = 0 \text{ to } 1). \tag{16}$$

In particular, if $\lambda = \mu$,

$$\left[\frac{\partial \pi_t(0)}{\partial t}\right] \mathrm{d}t = \frac{\lambda\,\mathrm{d}t}{(1+\lambda t)^2}, \qquad (t = 0 \text{ to } \infty). \tag{17}$$

From (16),

$$E\{t\} = -\frac{1}{\lambda}\log\left(1 - \frac{\lambda}{\mu}\right). \tag{18}$$

Number (N) *of new individuals per ancestor*
In discrete time,

$$\pi(z) = G(z\pi(z)), \tag{19}$$

whence

$$E\{N\} = \frac{m_1}{1 - m_1}, \qquad \sigma^2\{N\} = \frac{v_1}{(1-m_1)^3}, \tag{20}$$

where m_1 and v_1 mean and variance from $G(z)$ (cf. Nei, 1971b).
In continuous time,

$$\pi(z) = p + qz\pi^2(z),$$

where

$$p = \mu/(\lambda + \mu) \qquad (>\tfrac{1}{2}),$$

whence

$$E\{N\} = q/(p-q), \qquad \sigma^2\{N\} = pq/(p-q)^3 \tag{21}$$

(cf. Bartlett, 1966 § 5.22).

Use of (17) *as an approximation in population genetics*

TABLE I. Distribution of time to extinction (generations) of a neutral mutant

t	Simulation[a]	e^{-1}	λ $\frac{1}{2}$
1	84	73·6	100·0
2	25	24·9	25·0
3	23	16·7	15·0
4	10	12·0	10·0
5	5	9·0	7·1
6	6	7·0	5·4
7	3	5·7	4·2
8	6	4·6	3·3
9–10	4	7·1	5·0
11–20	14	17·1	11·4
21–30	4	6·7	4·2
31–50	6	5·9	3·6
51–100	5	4·7	2·9
over 100	5	5·0	2·9
Total	200	200	200
χ^2 (13 d.f.)		10·9	15·6

[a] Kimura and Ohta (1969b).

If alternatively (16) were used (to correspond more closely with finite population for which $P_0 < 1$), we could take $\mu/\lambda = P_0 = 1 - \frac{1}{2}N$ in neutral case.

III. Recurrent Epidemics Model and its Application to Measles

In (1), let $\mathbf{N}_t$ denote number S_t susceptibles and I_t infectives, then postulated epidemic model may be written

$$\frac{\partial \pi_t}{\partial t} = \lambda(w^2 - zw)\frac{\partial^2 \pi_t}{\partial z \partial w} + \mu(1 - w)\frac{\partial \pi_t}{\partial w} + v(z - 1)\pi_t + \varepsilon(w - 1)\pi_t \quad (22)$$

(cf. McKendrick, 1926; Bartlett, 1949).

As (pre-vaccination) model for measles, take $\varepsilon = 0$ (no immigration of infection, $\mu = \frac{1}{2}$ (incubation period about two weeks), $\rho = v\lambda/\mu \sim 1/68{\cdot}2$.

From (22), when $\varepsilon = 0$,

$$\frac{dx}{dt} = \lambda E\{IS\} - \mu x, \qquad \frac{dy}{dt} = -\lambda E\{IS\} + v, \quad (23)$$

where $x = E\{I\}$, $y = E\{S\}$.

Deterministic approximation: $E\{IS\} \sim xy$.

Linearizing approximation: $x = m(1 + \mu)$, $y = n(1 + v)$, where $m = v/\mu$, $n = \mu/\lambda$,

$$\frac{du}{dt} = \mu v(1 + u) \sim \mu v, \qquad \frac{dv}{dt} = -\rho(u + v + uv) \sim -\rho(u + v). \qquad (24)$$

Period $2\pi\{\rho\mu - \frac{1}{2}\rho^2\}^{-\frac{1}{2}} = 73{\cdot}7$ weeks.
Damping coefficient proportional to ρ.
Expected time to "fade-out" of infection (Bartlett, 1960)

$$T \sim \sqrt{(2\pi n)}\, e^{\frac{1}{2}(m+n/m)^2/n}/(\mu m), \qquad (n \gg m). \qquad (25)$$

Below critical size ($\varepsilon > 0$), period $\tau \sim 1/\rho + \sqrt{1/\rho\varepsilon}$.
Post-vaccination model (Griffiths, 1971):

$$v' = vf, \text{ (v influx of susceptibles, $f > 0$)},$$

$$m' = fm,\ n' = n,\ \rho' = f\rho.$$

Period $\sim 2\pi/\sqrt{\rho'\mu} = [2\pi/\sqrt{\rho\mu}]/\sqrt{f}$,

$$T' = T(fm, n) \sim T(m, n/f^2).$$

Critical community size multiplied by factor $1/f^2$.
Below critical size, $\tau' \sim \tau/f$ if $\varepsilon' = f\varepsilon$.
(Broadly speaking, reduction in susceptibles spreads out epidemics, but does not lessen intensity; Griffiths also notes that average age of onset will be increased.)

IV. Age Distributions

(i) For a uni-sex population $N_t(u)$ with birth- and death-rates $\lambda(u)$ and $\mu(u)$ dependent on age u, the deterministic age density $f_1(u, t)$ is the expected age-density $E\{dN(u, t)\}/dt$ in the complete stochastic formulation. For f_1 we have the continuity equation

$$\frac{\partial f_1}{\partial t} + \frac{\partial f_1}{\partial u} = -\mu(u) f_1, \qquad (0 < u < t), \qquad (26)$$

together with the "renewal" equation at $u = 0$,

$$f_1(0, t) = \int_0^\infty \lambda(u) f_1(u, t)\, du. \qquad (27)$$

These equations determine $f_1(u, t)$ from given initial conditions. In particular,

$$f_1(u, t) = f_1(0, t - u) \exp\left[-\int_0^u \mu(v)\, dv \right], \qquad (0 < u < t); \tag{28}$$

and if the population is not intrinsically decreasing, $f_1(0, t)$ has in general the asymptotic solution $C\, e^{\kappa t}$, where κ (≥ 1) is the dominant root of the equation in s,

$$\int_0^\infty e^{-st} \lambda(t)\, e^{-\int_0^t \mu(v)\, dv}\, dt = 1. \tag{29}$$

(For the more complicated case of two sexes, see, for example, Goodman, 1968, Bartlett, 1970.)

(ii) In the case of proliferating cell populations, denote the four mitotic phases G_1 (resting), S (DNA synthesis), G_2 (resting) and M (mitosis) by $i = 1, 2, 3, 4$ respectively. Then we may wish to consider either

$$N_i(u, t), \qquad u \text{ age from beginning of cycle,}$$

or

$$N_i(u_i', t), \qquad u_i' \text{ age from beginning of phase.}$$

For simplicity consider the case of zero death-rate. Then, if $r_1(u)$ are transition rates from phase i,

$$\begin{aligned} Df_1 &= -r_1(u) f_1, \qquad (u > 0), \\ Df_i &= -r_i(u) f_i + r_{i-1}(u) f_{i-1} \qquad (i \neq 1), \end{aligned} \tag{30}$$

where

$$D \equiv \frac{\partial}{\partial t} + \frac{\partial}{\partial u}.$$

Asymptotically, if $f_i(u, t) \sim g_i(u)\, e^{\kappa t}$,

$$g_1 = C_1\, e^{-\kappa u - \int_0^u r_1(v)\, dv},$$

etc. If we conveniently define

$$r_{1,i}(u) = r_i(u) g_i(u) / [g(u) + \ldots g_i(u)], \qquad i \neq 1,$$

then

$$\begin{aligned} g_1(u) + \ldots g_i(u) &= C_1\, e^{-\kappa u - \int_0^u r_{1,i}(v)\, dv} \\ &= C_1\, e^{-\kappa u} \{1 - \Phi_{1,i}(u)\}, \end{aligned} \tag{31}$$

say, where $C_1 = 2\kappa$ from relation

$$\int_0^\infty 2\,\mathrm{e}^{-\kappa u}\,\mathrm{d}\Phi_{1,4}(u) = 1.$$

Alternatively, for u_i',

$$D_i f_i' = -r_i(u_i') f_i', \qquad (u_i' > 0),$$

and if

$$f_i'(u_i', t) \sim g_i'(u_i')\,\mathrm{e}^{\kappa t},$$
$$g_i'(u_i') = C_i'\,\mathrm{e}^{-\kappa u_i'}\{1 - \Phi_i(u_i')\}. \qquad (32)$$

If

$$L_i(s) = \int_0^\infty \mathrm{e}^{-su}\,\mathrm{d}\Phi_i(u), \qquad C_i' = \kappa/[1 - L_i(\kappa)],$$

and fraction of $g(u)$ in phase i is

$$\int_0^\infty g_i(u)\,\mathrm{d}u = 2\{-L_{1,i}(\kappa) + L_{1,i-1}(\kappa)\}.$$

(If phase times are *independent*, then

$$L_{1,i}(s) = L_1(s)L_2(s)\ldots L_i(s).)$$

References and Short Bibliography

I. General

Bartlett, M. S. (1949). Some evolutionary stochastic processes. *J. R. Statist. Soc.* B **11**, 211–29.

Bartlett, M. S. (1966). "An Introduction to Stochastic Processes." (2nd Edn.) Cambridge Univ. Press, London.

Cox, D. R. and Miller, H. D. (1965). "The Theory of Stochastic Processes." Methuen, London.

Kolmogorov, A. (1931). Über die analytische Methoden in der Wahrscheinlichkeitsrechnung. *Math. Ann.* **104**, 415–58.

Levins, R. (1966). The strategy of model building in population biology. *Amer. Scientist* **54**, 421–31.

Maynard Smith, J. (1968). "Mathematical Ideas in Biology." Cambridge Univ. Press, London.

*Williamson, M. (1972). "The Analysis of Biological Populations." Edward Arnold, London.

Wit, C. T. de (1970). Dynamic concepts in biology. *In* "The use of Models in Agricultural and Biological Research." (Ed. J. G. W. Jones), 9–15.

II. Ecology and epidemiology

*Bailey, N. T. J. (1957). "The Mathematical Theory of Epidemics." Griffin, London.
Bartlett, M. S. (1960). "Stochastic Population Models in Ecology and Epidemiology." Methuen, London.
Gause, G. F. (1934)."The Struggle for Existence." Dover Publications, New York.
Griffiths, D. A. (1971). Epidemic models (D. Phil. Thesis, University of Oxford).
Lotka, A. J. (1926). "Elements of Physical Biology." Dover Publications, New York.
MacArthur, R. H. (1970). Species packing and competitive equilibrium for many species. *Theor. Pop. Biol.* **1**, 1–11.
*MacArthur, R. H. and Wilson, E. O. (1967). "The Theory of Island Biogeography." Princeton Univ. Press, U.S.A.
McKendrick, A. G. (1926). Applications of mathematics to medical problems. *Proc. Edin. Math. Soc.* **44**, 98–130.
*Patil, G. P., Pielou, E. C. and Waters, W. E. (1971). "Statistical Ecology." (3 vols.) Penn. State Univ. Press, U.S.A.
*Pielou, E. C. (1969). "An Introduction to Mathematical Ecology." Wiley, New York.
Volterra, V. (1926). Variazioni e fluttuazioni del numero d'individui in specie animali conviventi. *Mem. Acad. Lineci Roma*, **2**, 31–113.

III. Population genetics

*Cavalli-Sforza, L. L. and Bodmer, W. F. (1971). "The Genetics of Human Populations." Freeman, San Francisco.
Ewens, W. J. (1963). Diffusion equation and pseudo-distribution in genetics. *J. R. Statist. Soc.* B **25**, 405–12.
*Ewens, W. J. (1969). "Population Genetics." Methuen, London.
Fisher, R. A. (1930). "The Genetical Theory of Natural Selection." Oxford Univ. Press, London.
Haldane, J. B. S. (1924). A mathematical theory of natural and artificial selection. *Trans. Camb. Phil. Soc.* **23**, 19–41.
*Karlin, S. (1969). "Equilibrium behaviour of Population Genetic Models with Non-random Mating." Gordon and Breach, New York.
Kimura, M. (1964). Diffusion models in population genetics. *J. Appl. Prob.* **1**, 177–232.
Kimura, M. and Ohta, T. (1969a). The average number of generations until fixation of a mutant gene in a finite population. *Genetics* **61**, 763–71.
Kimura, M. and Ohta, T. (1969b). The average number of generations until extinction of an individual mutant gene in a finite population. *Genetics* **63**, 701–9.
*Kimura, M. and Ohta, T. (1971). "Theoretical Aspects of Population Genetics." Princeton Univ. Press.
Moran, P. A. P. (1962). "Statistical Processes of Evolutionary Theory." Oxford Univ. Press, London.
Nei, M. (1971a). Extinction time of deleterious mutant genes in large populations. *Theor. Pop. Biol.* **2**, 419–25.
Nei, M. (1971b). Total number of individuals affected by a single deleterious mutation in large populations. *Theor. Pop. Biol.* **2**, 426–30.
Wright, S. (1931). Evolution in Mendelian populations. *Genetics* **16**, 97–159.

IV. Demography, including cell kinetics

Barrett, J. C. (1966). A mathematical model of the mitotic cycle and its application to the interpretation of percentage labelled mitoses data. *J. Natn. Cancer Inst.*, **37**, 443–50.

BARTLETT, M. S. (1969). Distributions associated with cell populations. *Biometrika* **56**, 391–400.

BARTLETT, M. S. (1970). Age distributions. *Biometrics* **26**, 377–85.

*CLEAVER, J. E. (1967). "Thymidine Metabolism and Cell Kinetics." Elsevier, Amsterdam.

GOODMAN, L. A. (1968). Stochastic models for the population growth of the sexes. *Biometrika* **55**, 469–87.

KENDALL, D. G. (1949). Stochastic processes and population growth. *J. R. Statist. Soc.* B **11**, 230–64.

*KEYFITZ, N. (1968). "Introduction to the Mathematics of Population." Addison-Wesley, New York.

LESLIE, P. H. (1945). On the use of matrices in certain population mathematics. *Biometrika* **34**, 183–212.

*MACDONALD, P. D. M. (1970). Statistical inference from the fraction labelled mitoses curve. *Biometrika* **57**, 489–503.

POWELL, E. O. (1956). Growth rate and generation time of bacteria, with special reference to continuous culture. *J. Gen. Microbiol.* **15**, 492–511.

SISKEN, J. E. and MORASCA, L. (1965). Intrapopulation kinetics of the mitotic cycle. *J. Cell. Biol.* **25(2)**, 179–89.

* *Authors with useful further bibliographies.*

Some historical remarks and recollections on multivariate analysis

Reprinted from Sankhya: *The Indian Journal of Statistics Series B*
Vol. **36** Pt. 2, (1974) pp. 107–114

Introduction

The outburst of activity in multivariate analysis in the last decade has been obviously influenced to a large extent by the development of computers. In particular, the exploratory numerical investigations classifiable under the general title of cluster analysis are closely linked with the availability of computer algorithms. Such algorithms have clearly been developed in response to wide demand, and the empirical and *ad hoc* elements present in their design are being gradually reduced, or at least are becoming more appreciated. Recent surveys in this area (e.g. Cormack, Jardine and Sibson, 1971; Sibson 1972), and indeed in more 'classical' multivariate analysis (e.g. Rao, 1972), may perhaps excuse my decision not to attempt in these remarks on multivariate analysis to consider some of these newer aspects, with which I am, or have become, less familiar. Current workers in these fields are of course anxious to advance without excessive delay in marking and studying their route. Nevertheless, unfamiliarity with earlier work is sometimes revealed by their re-treading of routes already explored, and justifies those more closely involved in earlier developments recalling from time to time their salient features. Even when the results have not been overlooked (and with the comprchensive bibliography by Anderson, Das Gupta and Styan, 1972 or such recent books as Kshirsagar, 1972, there will be less excuse), it can still be of interest to recall some of the personal and historical details of these early investigations.

Rao (1964) has rightly drawn attention to Fisher's vital role in these first developments, though he would no doubt agree with the linking of the names of Hotelling and Mahalanobis with that of Fisher in the key developments in the 30's. Of course, no developments, however original, do not also have their association with still earlier work. If I restrict 'classical' *multivariate* analysis to the analysis of a vector set of quantitative 'dependent' variates, this permits the exclusion of the earlier multiple regression and correlation

---

*Invited paper for P. C. Mahalanobis Memorial Volume.

Opening address at the Royal Statistical Society Conference on *Multivariate Analysis and its Applications*, held in April, 1973 at Hull.

with which Karl Pearson and Udny Yule, as well as Fisher, had been well acquainted.

It does not exclude the Rothamsted work on the simultaneous distribution of normal sample variances and covariances (the Wishart 1928 distribution which generalized Fisher's 1915 discussion of the bivariate case in connection with his derivation of the correlation coefficient distribution), or on the sampling distribution of the multiple correlation coefficient (Fisher, 1928). Both these sampling distributions were very relevant to later multivariate sampling theory, although in the former case my factorization approach (Bartlett, 1933) to the distribution often enabled results to be obtained more readily.

A little later Hoteling (1931) initiated the development of the appropriate theoretical tests by his multivariate generalization of Student's t-test, to be followed by Wilks (1932) in further multivariate likelihood ratio extensions. This enabled me (Bartlett, 1934) to present a general multivariate regression theory, including the appropriate likelihood ratio criterion in the case of multivariate normality, with its exact method of testing where known, and its moments where not.

2. Key contributions of Fisher, Hotelling and Mahalanobis

It was at this stage that three contributions from Fisher, Hotelling and Mahalanobis occurred contemporaneously. Firstly, Fisher became concerned with the efficient utilization of multivariate measurements for discriminatory purposes; one or two papers (e.g. Barnard, 1935) which preceded Fisher's own (1936) paper on this topic acknowledged their authors' debt to Fisher's suggestions.

Mahalanobis (1936) was concerned with the concept of a generalized distance between populations, introducing the measure

$$(\Delta\boldsymbol{\mu})'\boldsymbol{V}^{-1}(\Delta\boldsymbol{\mu}),$$

where $\Delta\mu$ is the (column) vector of mean differences in the different quantitative variates, $(\Delta\mu)'$ the transposed row vector, and V the variance-covariance matrix, assumed the same for the two populations.

Hotelling (1936) was the first to examine the general *canonical* correlation analysis between two vector sets of variates, as distinct from the multiple regression or correlation of *one* dependent variate with several so-called 'independent' variates, or from the reciprocal regression problem of several dependent variates with one 'independent' variate (a regression problem which included Fisher's linear discriminant function analysis if the 'independent' variate was appropriately defined as a dummy variate associated with the contrast between two groups).

At the time of my 1934 general formulation of the analysis of variance and covariance Hotelling's canonical reduction was thus not available, but its publication in 1936 stimulated me to consider how far my technique for deriving a χ^2 approximation to likelihood ratio criteria valid to the *second* order of approximation, which I applied in 1938 to the multivariate regression problem, could be used in the *sample* canonical factorization of the likelihood criterion, as distinct from a direct factorization on *a priori* variates.

3. Value of the geometrical approach

It was at this stage that I started to have some criticisms of the way Fisher was presenting multivariate *sampling* theory. He himself appeared to be abandoning his theoretical geometrical approach in favour of arithmetical or algebraic analogies which at times led him into error (cf. Bartlett, 1965b, p. 407). Thus my own guarded comments (1938, Section 9) on significance testing for *general* canonical analysis were in contrast with his own original version in his 1938 paper. (See end of Section 5, p. 383, which was in fact wrong in its allocation of degrees of freedom, and corrected in a typescript erratum, although still uncritically, after I had sent him a copy of my own paper !) Furthermore, my notes (1939a and b) on *exact* tests of significance for discriminant function variables taken separately were not in agreement with Fisher's original recommendation (1938; cf. also, however, 1940).

The 1939-45 war tended to slow down the developments at this point, though not before the brilliant theoretical derivations by Fisher (1939) and Hsu (1939) of the simultaneous null distribution of the sample canonical correlations. My own paper on the non-null distribution (1947a) was in the spirit of Fisher's earlier (1928) paper on the non-null distribution of the multiple correlation coefficient. As I have previously remarked (1965b), before the war I did not understand Fisher's paper, and preferred to use Wilks' analytical derivation in lectures; only later did an understanding of Fisher's derivation allow me to progress with the canonical correlation distribution.

Such developments, however, were not of immediate use in the canonical breakdown' of multivariate analyses, and I still used my approximate χ^2 analyses (cf. Bartlett, 1947b) in practice, in spite of the additional uncertainty when applied to sample *canonical* roots) mentioned above. Later results by D. N. Lawley (1956) enabled somewhat improved approximations in this more dubious situation to be made, but a valuable alternative technique first discovered by E. J. Williams (1952) in particular cases was generalize by me in a paper which, somewhat embarrassingly, appeared first. Thi technique consisted of factorizing out *hypothetical* canonical variables in plac of sample ones, enabling exact factorizations (in the sense of factors wit known moments) to replace approximate ones. The use of the χ^2 approxi mation then became a matter of convenience rather than of necessity, and i fact later no longer was needed following the publication of Schatzoff's table (1966). The technique had the dual value of (i) permitting a check on th χ^2 approximation tentative analyses (ii) enabling more precise confidenc intervals and regions for the sample canonical variates to be explored in th case of more than one 'independent' variable, one independent (dummy variable being the discriminant function case. I have often felt that thes factorization techniques have been exploited somewhat less often than the could have been by workers other than Williams and myself (see, however Kshirsagar, 1972); one still finds statisticians contenting themselves with a over-all test of significance instead of exploring the structure of the relation in greater detail.

The discriminatory relations of several groups or populations, which is in cluded as a particular application of the generalized regression theory (Bartlett 1951), has alternatively been studied by some workers (e.g. Rao, 1952) b means of the generalized distance between every pair of groups.

4. Recapitulation of exact Λ factorizations

In view of their wide applicability, let me summarize the various 'exac factorizations available, expressed in their most convenient algebraic forn (completed by Kshirsagar, 1964, 1971). We express the n sample values o each of p 'dependent' variables by the matrix Y (p rows) and of the q 'indepen dent' variables by X. More generally, we assume that a set of 'nuisanc variables Z has already been eliminated by linear regression (including elim nation of group or block constants by the use of dummy variables, of whic the most common is the unit vector used to eliminate the general mean), an Y and X are the residual deviations with n denoting the appropriate numbe of degrees of freedom. X may also of course represent dummy variable as in discrimination problems. We denote the matrix products YY', YX XX', where dashes denote transposes by C_{YY}, $C_{YX} = C'_{XY}$, C_{XX}, and writ

$$C_{YY\cdot X} = C_{YY} - C_{YX}C_{XX}^{-1}C_{XY}.$$

Then we use Wilks' criterion

$$\Lambda = |C_{YY\cdot X}|/|C_{YY}|,$$

where $|C_{YY}|$ denotes the determinant of C_{YY}, as our overall test statistic, and write it more fully as $\Lambda(n, p, q)$.

Consider now two alternative situations (a) and (b). In (a) we suppose that $\beta' Y$ is a hypothetical linear combination of the p dependent variates exhausting the relations with X. It will consequently be the *true* first canonical variate, the others being null. (This case can be generalized, but we confine our attention to it.) There will be a corresponding true first canonical variate of the 'independent' variates, represented in the sample by, say, $\alpha' X$. In (b) we consider the alternative situation where $\alpha' X$ is first given, rather than $\beta' Y$.

In either case, as there is only one non-zero true canonical variate, the elimination of this variate should effectively remove the relationship between Y and X. This leads to *two* possible factorizations viz. in case (a)

(i) $\Lambda(n, p, q) = \Lambda(n, 1, q)\Lambda(n-1, p-1, 1)\Lambda(n-2, p-1, q-1)$

(ii) $\Lambda(n, p, q) = \Lambda(n, 1, q)\Lambda(n-1, p-1, q-1)\Lambda(n-q, p-1, 1)$.

The first factor on the right is the same for (i) and (ii) corresponding to the elimination of the (linear) dependence of $\beta' Y$ on X. If we write $C_{YY \cdot X} = A$ and $C_{YY} = A+B$, then

$$\Lambda(n, 1, q) = \frac{\beta' A \beta}{\beta'(A+B)\beta},$$

leaving the factor $\Lambda(n-1, p-1, q)$. In (i) this has been factored into a 'direction' factor, which in effect tests the discrepancy between $\beta' Y$ and the sample first canonical variate $b' Y$, say; and the 'collinearity' factor 'partial' to the direction factor, representing the residual measure of dependence between Y and X. The first factor is

$$\Lambda(n-1, p-1, 1) = \left(1 - \frac{\beta' B(A+B)^{-1} B\beta}{\beta' B\beta}\right) / \Lambda(n, 1, q)$$

and the second is then obtainable from (i) above. The alternative factorization (ii) is perhaps less natural, corresponding to the collinearity factor $\Lambda(n-1, p-1, q-1)$ and the partial direction factor $\Lambda(n-q, p-1, 1)$. Here

$$\Lambda(n-1, p-1, q-1) = \Lambda(n, p, q)\left(1 + \frac{\beta' B A^{-1} B\beta}{\beta' B\beta}\right)$$

and the remaining factor is obtainable from (ii).

In case (b) the only change in equations (i) and (ii) is the interchange of p and q. For their algebraic expression, denote $\alpha' X$ by z. Then for the new factorization (i) we have the direction factor

$$\Lambda(n-1, q-1, 1) = \frac{\gamma' A \gamma}{\gamma' C_{Yz}} \frac{C_{zz}}{C_{zz} - \gamma' C_{Yz}},$$

where $\gamma = (A+B)^{-1}C_{Yz}$, and the partial collinearity factor is obtainable from (i). For the new factorization (ii) we have the collinearity factor

$$\Lambda(n-1, p-1, q-1) = \Lambda(n, p, q)\, \frac{C_{zY}A^{-1}C_{Yz}}{\gamma' C_{Yz}}$$

and the partial direction factor is obtainable from the new (ii).

In all this exact sampling theory, the minimum assumption for validity of the tests is the joint (residual) normality (with constant dispersion matrix over the values of the other set) of *one* of the sets Y or X; though even without the normality assumption useful results may still be possible (cf. Bartlett, 1951).

Of course the use of a hypothetical function on which to hinge the analysis is less justified if this function is incompatible with the data, so that some caution in the interpretation of the collinearity factor obtained by one of these exact factorizations would then be advisable. The direct but 'approximate' factor referred to earlier, obtained by eliminating the first *sample* canonical variable, does not have such a limitation.

To illustrate this, let me refer to the example first used in my 1934 paper, with only two variates y_1 (grain) and y_2 (straw) related to eight manurial treatments ($q = 7$), in an experimental layout in nine blocks, block means being eliminated ($n = 7\times 8 = 56$). The relevant sums of squares and products, writing them in rows for convenience rather than as matrices, were :

	y_1^2	y_1y_2	y_2^2	*D.F.*
Treatments	32,985.0	− 6,786.6	12,496.8	7
Residual	71,496.1	58,549.0	136,972.6	49
Total	104,481.1	51,762.4	149,469.4	56

From these figures we have (see Bartlett, 1951) $\Lambda = 0.4920$, $r_1^2 = 0.47698$ and $r_2^2 = 0.05934$, with the first (and only significant) sample canonical variate $y_1 - 0.535y_2$ (95% confidence limits -0.827 to -0.275).

Clearly the hypothetical function y_1+y_2 (total yield) is quite unacceptable. Nevertheless, *if we persist in using it*, the factorizations of Λ are given by

$$\Lambda(56, 2, 7) = \Lambda(56, 1, 7)\Lambda(55, 1, 1)\, \Lambda(54, 1, 6)$$
$$0.4920 = (0.91074)(0.8573)(0.6302)$$

or

$$\Lambda(56, 2, 7) = \Lambda(56, 1, 7)\Lambda(55, 1, 6)\Lambda(49, 1, 1)$$
$$0.4920 = (0.91074)(0.7911)(0.6829)$$

with χ^2 analyses of the last two factors (stars denote degree of significance)

Direction	8.24** (1 d.f.)	Partial direction	11.95*** (1 d.f.)
Partial collinearity	23.09*** (6 d.f.)	Collinearity	18.12** (6 d.f.)
Total (non-additive)	31.10*** (7 d.f.).		

The factorization of the over-all test of the inadequacy of y_1+y_2 is meaningless in either case, but the approximate χ^2 test of non-collinearity based on r_2^2 gives a non-significant 3.19 (6 d.f.; this figure differs slightly from the 3.12 given in Bartlett, 1956, due to the incorporation of Lawley's later correction). It is of interest, but would therefore be rather misleading, to note that these further exact factorizations are (as is clear from Kshirsagar's formulation of them) not dependent on the *sample* canonical correlations, and might therefore chronologically have preceded, rather than have followed, Hotelling's canonical analysis.

References

Anderson, T. W., Das Gupta, S and Styan, G. P. H. (1972): *A Bibliography of Multivariate Statistical Analysis*, Oliver & Boyd, Edinburgh.

Ashton, E. H., Healy, M. J. R. and Lipton, S. (1957): The descriptive use of discriminant functions in physical anthropology. *Proc. Roy. Soc.*, B, **146,** 552.

Barnard, M. M. (1935): The secular varieties of skull characters in four series of Egyptian skulls. *Ann. Eugen.* 6, 352.

Bartlett, M. S. (1933): On the theory of statistical regression. *Proc. Roy. Soc. Edin.*, **53,** 260.

——— (1934): The vector representation of a sample. *Proc. Camb. Phil. Soc.*, **30,** 327.

——— (1938): Further aspects of the theory of multiple regression. *Proc. Camb. Phil. Soc.*, **34,** 33.

——— (1939a): A note on tests of significance in multivariate analysis. *Proc. Camb. Phil. Soc.*, **35,** 180.

——— (1939b): The standard errors of discriminant function coefficients. *J. R. Statist. Soc.*, (Suppl.) **6,** 169.

——— (1947a): The general canonical correlation distribution. *Ann. Math. Statist.*, **18,** 1.

——— (1947b): Multivariate analysis. *J. R. Statist. Soc.*, (Suppl.), **9,** 176.

——— (1951): The goodness of fit of a single hypothetical discriminant function in the case of several groups. *Ann. Eugen.*, **16,** 199.

——— (1965a): Multivariate statistics. *Theoret. and Math. Biology*, Blaisdell, N. York Ch. 8.

——— (1965b): R. A. Fisher and the last fifty years of statistical methodology. *J. Amer. Statist. Soc.*, **60,** 395.

Cormack, R. M. (1971): A review of classification. *J. R. Statist. Soc.*, B, **134,** 321.

Fisher, R. A. (1928): The general sampling distribution of the multiple correlation coefficient. *Proc. Roy. Soc.*, A **121,** 654.

——— (1936): The use of multiple measurements in taxonomic problems. *Ann. Eugen.*, **7,** 179.

——— (1938): The statistical utilization of multiple measurements. *Ann. Eugen.*, **8,** 376.

FISHER, R. A. (1939): The sampling distribution of some statistics obtained from non-linear equations. *Ann. Eugen.*, **9**, 238.

——— (1940): The precision of discriminant functions. *Ann. Eugen.*, **10**, 422.

——— (1946): *Statistical Methods for Research Workers*, Oliver & Boyd, Edin., 10th ed.

HOTELLING, H. (1931): The generalization of Student's ratio. *Ann. Math. Statist.*, **2**, 360.

——— (1933). Analysis of a complex of statistical variables into principal components. *J. Educ. Psychol.*, **24**, 417 & 498.

——— (1936): Relations between two sets of variates. *Biometrika*, **28**, 321.

HSU, P. L. (1939): On the distribution of roots of certain determinantal equations. *Ann. Eugen.*, **9**, 250.

JARDINE, N. and SIBSON, R. (1971): *Mathematical Taxonomy* (Wiley, N. York).

KSHIRSAGAR, A. M. (1962): A note on direction and collinearity factors in canonical analysis. *Biometrika*, **49**, 255.

——— (1964). Distribution of the direction and collinearity factors in discriminant analysis. *Proc. Camb. Phil. Soc.*, **60**, 217.

——— (1971): Goodness of fit of a discriminant function from the vector space of dummy variables. *J. R. Statist. Soc.*, B **33**, 111.

——— (1972): *Multivariate Analysis* (M. Dekker, N. York).

LAWLEY, D. N. (1956). A general method for approximating to the distribution of likelihood ratio criteria. *Biometrika*, **43**, 295.

MAHALANOBIS, P. C. (1936): On the generalized distance in statistics. *Proc. Nat. Inst. Sci. India*, Pt. A, **2**, 49.

PEARSON, E. S. and WILKS, S. S. (1933): Methods of statistical analysis appropriate for k samples of two variables. *Biometrika*, **25**, 353.

PEARSON, E. S. and HARTLEY, H. O. (ed.) (1972): *Biometrika Tables for Statisticians*, **2**, (Cambridge Univ. Press).

RAO, C. R. (1948): The utilization of multiple measurements in problems of biological classification. *J. R. Statist. Soc.*, B, **10**, 159.

——— (1952): *Advanced Statistical Methods in Biometric Research* (Wiley, N York).

——— (1964): Sir Ronald Aylmer Fisher—the architect of multivariate analysis. *Biometrics*, **20**, 286.

——— (1972): Recent trends of research work in multivariate analysis. *Biometrics*, **28**, 3.

SCHATZOFF, M. (1966): Exact distributions of Wilk's likelihood ratio criterion. *Biometrika*, **53**, 347.

SEAL, H. (1964): *Multivariate Statistical Analysis for Biologists* (Methuen, Lond.).

SIBSON, R. (1972): Order invariant methods for data analysis. *J. R. Statist. Soc.*, B **34**, 311.

SMITH, H. F. (1936): A discriminant function for plant selection. *Ann. Eugen.*, **7**, 240.

WILKS, S. S. (1932): Certain generalizations in analysis of variance. *Biometrika*, **24**, 471.

WILLIAMS, E. J. (1952): Some exact tests in multivariate analysis. *Biometrika*, **39**, 17.

——— (1959): *Regression Analysis* (Wiley, N. York).

WISHART, J. (1928): The generalized product moment distribution in samples from a normal multivariate population. *Biometrika*, **20A**, 32.

WISHART, J. and BARTLETT, M. S. (1933): Generalized product moment distribution in a normal system. *Proc. Camb. Phil. Soc.*, **29**, 260.

Paper received : November, 1973.